London Mathematical Society Lecture Note Series. 13

Combinatorics

The Proceedings of the British Combinatorial Conference held in the University College of Wales, Aberystwyth, 2-6 July 1973

N. Biggs, P. J. Cameron, D. V. Chopra, R. L. Constable, R. J. Cook, F. Escalante, S. Fiorini and R. J. Wilson, A. Gardiner, R. K. Guy A. Hill, A. J. W. Hilton, D. A. Holton, R. P. Jones, A. D. Keedwell, E. K. Lloyd, L. Lovász and M. D. Plummer, J. C. P. Miller, J. W. Moon, R. Nelson, J. Schonheim, J. Sheehan, D. H. Smith, H. N. V. Temperley, W. T. Tutte, J. H. van Lint, D. A. Waller, R. J. Wilson, D. R. Woodall.

Edited by
T. P. McDONOUGH &
V. C. MAVRON

Cambridge · At the University Press · 1974

CAMBRIDGE UNIVERSITY PRESS
Cambridge, New York, Melbourne, Madrid, Cape Town, Singapore, São Paulo

Cambridge University Press
The Edinburgh Building, Cambridge CB2 8RU, UK

Published in the United States of America by Cambridge University Press, New York

www.cambridge.org
Information on this title: www.cambridge.org/9780521204545

First published 1974
Re-issued in this digitally printed version 2008

A catalogue record for this publication is available from the British Library

Library of Congress Catalogue Card Number: 74–76571

ISBN 978-0-521-20454-5 paperback

Contents

Preface

The British Combinatorial Conference, 2nd-6th July 1973, held at Aberystwyth, was the fourth such conference held in Britain in the past decade.

Most lectures were of twenty to forty minutes duration. There were, however, four longer lectures from invited speakers. These four were H. Lüneburg (Kaiserslautern), C. St. J. A. Nash-Williams (Aberdeen), W. T. Tutte (Waterloo) and J. H. van Lint (Eindhoven).

It was envisaged that this conference should be one of a regular series, aimed at providing a forum for combinatorial mathematicians in Britain. It was also hoped that there would be considerable participation by mathematicians from outside Britain. That this hope was fulfilled is seen from the substantial number (thirty per cent) of such mathematicians attending.

We would like to thank the University College of Wales, and in particular its Department of Pure Mathematics, for the assistance given to the organisers both prior to, and during, the conference. We also wish to acknowledge, with gratitude, the financial assistance given by the Royal Society.

16. 10. 1973

T. P. McDonough
V. C. Mavron

PERFECT CODES AND DISTANCE-TRANSITIVE GRAPHS

NORMAN BIGGS

1. Introduction

Let S_k denote the set of sequences of k binary digits; in coding theory a subset C of S_k is called a binary code of block length k. If a code-word $c \in C$ is 'transmitted', and a sequence $s \in S_k$ is 'received', then the number of errors is the number of places in which s differs from c. One defines

$$\Sigma_e(c) = \{s \in S_k \mid s \text{ and } c \text{ differ in at most } e \text{ places}\},$$

and says that C is an e-error correcting code if the sets $\Sigma_e(c)$, as c runs through C, are disjoint. If these sets partition S_k, we have a perfect code.

In coding theory it is customary to introduce the vector space structure of the set S_k; however, we shall take the view that the elements of S_k are best regarded as the vertices of a graph, two vertices being adjacent whenever they differ in just one place. We denote this graph by the symbol Q_k, and note that it is the graph formed by the vertices and edges of a hypercube in k dimensions. The distance function ∂ in Q_k enables us to count errors, and we now write

$$\Sigma_e(v) = \{w \in VQ_k \mid \partial(v, w) \leq e\}.$$

In these terms, an e-error correcting binary perfect code, of block length k, is a subset C of VQ_k with the property that the sets $\Sigma_e(c)$, as c runs through C, partition VQ_k. We shall refer to C as a perfect e-code in Q_k, and we shall always take $e \geq 1$.

It is remarkable that there are relatively few pairs (k, e) for which a perfect e-code in Q_k exists [7], [8]. The complete list is:

(i) $k = e$, the trivial codes with $|C| = 1$;

(ii) $k = 2e + 1$, the 'repetition' codes with $|C| = 2$;

(iii) $k = 2^r - 1$, $e = 1$, the Hamming codes [8];

(iv) $k = 23$, $e = 3$, the binary Golay code [8].

We are led to consider the possibility of replacing Q_k by other graphs. If Γ is a finite, connected, simple graph, with distance function ∂, and the sets $\Sigma_e(v)$ are defined as for Q_k, then we say that a subset C of $V\Gamma$ is a perfect e-code in Γ if the sets $\Sigma_e(c)$, $c \in C$, partition $V\Gamma$.

Now it is clear that for any given $e \geq 1$ we can construct, at will, graphs Γ which possess perfect e-codes, for we may just take a set of neighbourhoods $\Sigma_e(c)$ and join their free ends by extra edges; however, the graphs so constructed are uninteresting. We claim that the natural setting for the problem of perfect codes is the class of distance-transitive graphs [2]. This claim will be justified in Section 3, after some motivation in Section 2.

2. Perfect 1-codes in regular graphs

Suppose that Γ is regular, with valency k, and let A denote its adjacency matrix. If $\mathbf{c}$ is the column vector whose entries are 1 in positions corresponding to the vertices of a perfect 1-code in Γ, and 0 elsewhere, then

$$A\mathbf{c} = \mathbf{u} - \mathbf{c}$$

where $\mathbf{u}$ is the vector each of whose entries is 1. Let

$$\mathbf{w} = \mathbf{u} - (k + 1)\mathbf{c}.$$

Then we have

$$A\mathbf{w} = A\mathbf{u} - (k + 1)A\mathbf{c} = k\mathbf{u} - (k + 1)(\mathbf{u} - \mathbf{c}) = -\mathbf{w}.$$

In other words, -1 is an eigenvalue of A, corresponding to the eigenvector $\mathbf{w}$. Since A is a rational symmetric matrix, its minimum polynomial $\mu(t)$ belongs to the ring $\mathbf{Q}[t]$ of polynomials with rational coefficients. We call $\mu(t)$ the minimum polynomial of Γ, and we have proved:

Theorem 1. If the regular graph Γ has a perfect 1-code, then $t + 1$ is a divisor of $\mu(t)$ in the ring $\mathbf{Q}[t]$.

The result indicates that the minimum polynomial of a graph is relevant to the study of perfect codes in the graph. In the case of a distance-transitive graph, not only do we have a simple method of finding the minimum polynomial, but there is also an extension of Theorem 1 for perfect e-codes with $e > 1$.

3. Perfect e-codes in distance transitive graphs

The graph Γ is distance-transitive if whenever u, v, x, y are vertices of Γ such that $\partial(u, v) = \partial(x, y)$ then there is an automorphism of Γ taking u to x and v to y. A full treatment of the properties of such graphs may be found in [2], but we shall sketch the relevant parts of the theory here.

Associated with each distance-transitive graph Γ, having valency k and diameter d, is an intersection array

$$\iota(\Gamma) = \left\{ \begin{matrix} * & 1 & c_2 & \cdots & c_{d-1} & c_d \\ 0 & a_1 & a_2 & \cdots & a_{d-1} & a_d \\ k & b_1 & b_2 & \cdots & b_{d-1} & * \end{matrix} \right\};$$

from this we can calculate the eigenvector sequence $v_0(t), v_1(t), \ldots, v_d(t)$, each term of which belongs to the ring $\mathbf{Q}[t]$. The recursion defining this sequence is

$$\begin{cases} v_0(t) = 1, \quad v_1(t) = t, \\ c_i v_i(t) + (a_{i-1} - t)v_{i-1}(t) + b_{i-2}v_{i-2}(t) = 0 \quad (i = 2, \ldots, d). \end{cases}$$

For $0 \le i \le d$ we define $x_i(t) = v_0(t) + v_1(t) + \ldots + v_i(t)$; then it can be shown that the minimum polynomial of Γ is

$$\mu(t) = (t - k)x_d(t).$$

The proof of the following theorem is given in [1].

Theorem 2. If the distance-transitive graph Γ has a perfect e-code, then $x_e(t)$ is a divisor of $\mu(t)$ in the ring $\mathbf{Q}[t]$.

We notice that $x_1(t) = t + 1$, so that we have verified incidentally the result of Theorem 1 in this special case.

The graph Q_k is a distance-transitive graph, with intersection array

$$\iota(Q_k) = \left\{ \begin{matrix} * & 1 & 2 & . & . & . & k-1 & k \\ 0 & 0 & 0 & . & . & . & 0 & 0 \\ k & k-1 & k-2 & . & . & . & 1 & * \end{matrix} \right\}$$

Now it follows from [1, Section 5] that, if we write $s = \frac{1}{2}(k - t)$, then

(i) $$x_e(t) = \sum_{i=0}^{e} (-1)^i \binom{s-1}{i} \binom{k-s}{e-i},$$

(ii) $$\mu(t) = Rs(s-1)(s-2) \dots (s-k) \quad (R \text{ a rational constant}).$$

We deduce from Theorem 2 that if there is a perfect e-code in Q_k, then the polynomial on the right of (i) must have its e zeros corresponding to s in the set $\{0, 1, \dots, k\}$. This is the theorem of Lloyd [8], in the classical case, and it was by using this theorem that the list in Section 1 was proved to be complete.

It is now possible to state three reasons why the question of perfect codes should be considered in the context of distance-transitive graphs.

(a) The classical question is a special case.

(b) The theorem of Lloyd generalizes and simplifies.

(c) Other interesting examples arise.

4. Examples

Examples of perfect codes in distance-transitive graphs are rare; in fact, it is true to say that examples of distance-transitive graphs are rare! However, this merely adds interest to the examples which are known.

It is clear that the graphs Q_k can be generalized by replacing the binary 'alphabet' by an alphabet of q symbols, for any $q > 2$. This generalization is part of classical coding theory, and is treated from our present viewpoint in [1]. It is known that, apart from some perfect 1-codes, the only other code in this case is the ternary Golay 2-code [8].

In the twelve trivalent distance-transitive graphs [4] there are only two non-trivial perfect codes: the repetition 1-code in Q_3 and a 1-code in the graph with 28 vertices. The latter code is evident from the construction of the graph given in Section 1 of [4].

We now turn to the odd graphs O_k $(k \geq 3)$. The graph O_k has for its vertices the (k-1)-subsets of a (2k-1)-set, two vertices being adjacent whenever the subsets are disjoint; O_k is a distance-transitive graph with valency k and diameter $k - 1$. It can be shown that the eigenvalues of O_k are the integers $(-1)^{k-i}i$ $(1 \leq i \leq k)$, so that

$$\mu(t) = (t - k)(t + k - 1)(t - k + 2) \dots (t + (-1)^k).$$

It is also easy to give explicit expressions for the first few terms of the eigenvector sequence, and from these we find

$$x_0(t) = 1, \quad x_1(t) = t + 1, \quad x_2(t) = t^2 + t - (k - 1),$$
$$x_3(t) = \tfrac{1}{2}(t + 1)(t^2 + t - (2k - 2)).$$

Theorem 3. Suppose that there is a perfect e-code in O_k, $(e = 1, 2, 3)$. Then

(i) $e = 1 \Rightarrow k$ is even;
(ii) $e = 2 \Rightarrow k = 4r^2 - 2r + 1$ for some natural number r;
(iii) $e = 3 \Rightarrow k = 2(4r^2 - 3r + 1)$ for some natural number r.

Proof. (i) For a 1-code in O_k we require that $t + 1$ is a factor of $\mu(t)$, and this is so if and only if k is even.

(ii) For a 2-code in O_k we require that $t^2 + t - (k - 1)$ divides $\mu(t)$. Since the zeros of $\mu(t)$ are the integers $(-1)^{k-i}i$ $(1 \leq i \leq k)$ we must have

$$t^2 + t - (k - 1) = (t - \alpha)(t - \beta),$$

where α and β are integers having the stated form. Equating co-efficients of t we get $\beta = -(\alpha + 1)$, and we may assume that $\alpha > 0$,

$\beta < 0$. Equating coefficients of unity we get

$$k - 1 = -\alpha\beta = \alpha(\alpha + 1),$$

so that $k - 1$ is even and k is odd. Since α is a positive integral zero of $\mu(t)$, $k - \alpha$ must be even, and so α is odd. Writing $\alpha = 2r - 1$, we get $k = 2r(2r - 1) + 1 = 4r^2 - 2r + 1$, as required.

(iii) This part is proved by an argument like that in (ii).

Our condition that k is even for a 1-code in O_k is a weak one, and it can be improved by the following direct argument. Let C be a subset of VO_k which is a perfect 1-code; then any two distinct vertices u, v in C satisfy $\partial(u, v) \geq 3$. But if these vertices (regarded as $(k-1)$-subsets of a $(2k-1)$-set) have $k - 2$ elements in common, then $\partial(u, v) = 2$. Consequently each set of $k - 2$ elements occurs at most once as a subset of the elements in a vertex belonging to C. Since each vertex contains $k - 1$ sets of $k - 2$ elements we have

$$|C| \leq \frac{1}{k-1} \cdot \binom{2k-1}{k-2}$$

with equality only if each $(k-2)$-set occurs exactly once in a vertex of C. But for a perfect 1-code, the $\binom{2k-1}{k-1}$ vertices are partitioned into $|C|$ sets of $k + 1$, and so

$$|C| = \frac{1}{k+1} \cdot \binom{2k-1}{k-1} = \frac{1}{k-1} \cdot \binom{2k-1}{k-2}.$$

Thus every $(k-2)$-set occurs just once in a vertex of C, and these vertices are the blocks of a Steiner system $S(k-2, k-1, 2k-1)$. (This result is due to P. J. Cameron.) There are only two such systems known: $S(2, 3, 7)$ and $S(4, 5, 11)$, giving rise to perfect 1-codes in O_4 and O_6. In fact the divisibility conditions for a Steiner system imply that $k + 1$ must be prime, which is considerably stronger than our requirement that $k + 1$ must be odd.

There are no known e-codes in O_k for $k-1 > e > 1$.

We now mention a situation which generalizes the 'repetition' codes in the classical case. We say that a connected graph Γ, of diameter d, is antipodal if $\partial(u, v) = d$ and $\partial(u, w) = d$ implies that $v = w$ or

$\partial(v, w) = d$. The importance of this concept lies in the fact that a distance-transitive graph in which the automorphism group acts imprimitively on the vertices must be either bipartite or antipodal [6]. An antipodal distance-transitive graph Γ of odd diameter $d = 2d' + 1$ has a derived graph Γ', with diameter d', which is also distance-transitive; details of this situation are given in [3]. We find that $|V\Gamma| = r|V\Gamma'|$ for some integer $r \geq 2$, and Γ has a perfect d'-code C with $|C| = r$. Furthermore, the calculations in [3] show that, for Γ, $x_{d'}(t)$ divides $\mu(t)$, in accordance with Theorem 2.

Finally, we construct a special example. Consider the projective plane $PG(2, 3^2)$; this plane admits a unitary polarity induced by the field automorphism $\theta \mapsto \theta^3$ of $GF(3^2)$. The plane contains 91 points and 91 lines, which may be classified as follows [5]:

28 isotropic points	(points which lie on their polar lines);
63 non-isotropic points	(points which do not lie on their polar lines);
28 tangents	(lines containing 1 isotropic point and 9 non-isotropic points);
63 secants	(lines containing 4 isotropic points and 6 non-isotropic points).

We construct a graph W, whose vertices are the 63 non-isotropic points, and two are adjacent whenever each lies on the polar line of the other. Then W is a distance-transitive graph with intersection array

$$\begin{Bmatrix} * & 1 & 1 & 3 \\ 0 & 1 & 1 & 3 \\ 6 & 4 & 4 & * \end{Bmatrix}$$

and minimum polynomial

$$(t - 6)(t + 1)(t^2 - 9).$$

The graph W has a perfect 1-code, consisting of the 9 vertices corresponding to the non-isotropic points on any tangent.

References

1. N. L. Biggs. Perfect codes in graphs, J. Combinatorial Theory (B), 15 (1973), 289-96.
2. N. L. Biggs. Algebraic graph theory. Cambridge Tracts in Mathematics, 67, Cambridge University Press, London, 1974.
3. N. L. Biggs and A. D. Gardiner. On antipodal graphs. (In preparation.)
4. N. L. Biggs and D. H. Smith. On trivalent graphs, Bull. London Math. Soc., 3 (1971), 155-8.
5. P. Dembowski. Finite geometries, Springer, Berlin, 1968.
6. D. H. Smith. Primitive and imprimitive graphs, Quart. J. Math. (Oxf.), 22 (1971), 551-7.
7. A. Tietavainen. On the non-existence of perfect codes over finite fields, Siam J. Appl. Math., 24 (1973), 88-96.
8. J. H. van Lint. Coding theory. Lecture Notes in Mathematics, 201, Springer, Berlin, 1971.

Royal Holloway College,
London, England

GENERALISATION OF FISHER'S INEQUALITY TO FIELDS WITH MORE THAN ONE ELEMENT

PETER J. CAMERON

Many people (Petrenjuk, Wilson, Ray-Chaudhuri, Noda, Bannai, Delsarte, Goethals, and Seidel among them) have contributed to these results; some of the ideas arose in several places. So this article will tend to be a commentary on the facts. I define a t-design, with parameters v, k, b_t, to be a collection of k-subsets of the v-set X, called 'blocks', with the property that any t-subset is contained in precisely b_t blocks; I require the non-degeneracy condition $t \leq k \leq v-t$. A t-design is a t'-design for $0 \leq t' \leq t$. I shall use b for the number of blocks, though notation suggests b_0. Fisher's inequality states that, in a 2-design, $b \geq v$; furthermore, if equality holds, then the 2-design is called symmetric, and has the property that the size of the intersection of two blocks is constant. The generalisations I shall discuss are:

(1) In a 2s-design, $b \geq \binom{v}{s}$; if equality holds, then for distinct blocks B, B', $|B \cap B'|$ takes just s distinct values.

(2) In a (2s-2)-design in which, for distinct blocks B, B', $|B \cap B'|$ takes just s distinct values, $b \leq \binom{v}{s}$.

(In (2) it is also true that the blocks carry an 'association scheme with s classes', defined in the obvious way.)

If the definition of a t-design is weakened to allow 'repeated blocks', (1) remains true, while the only counterexamples to (2) are obtained by taking a (2s-2)-design without repeated blocks in which $|B \cap B'|$ takes just $s - 1$ values (such a design has exactly $\binom{v}{s-1}$ blocks), and repeating each block the same number of times.

The only known examples of equality in (1) with $s \geq 2$ are the Steiner system $S(4, 7, 23)$ (a 4-design with $v = 23$, $k = 7$, $b_4 = 1$) and its complement.

(1) is clearly a generalisation of Fisher's inequality; (2) is slightly less obviously so - we must observe that the 'dual' of the case $s = 1$ of

(2) is the case $s = 1$ of the following strengthened version of the first part of (1):

(3) Let $\mathcal{B}$ be a collection of subsets of a set X with $|X| = v$, and s an integer, such that

(i) for $s \le i \le 2s$, the number of members of $\mathcal{B}$ containing an i-subset of X is a constant b_i, depending only on i;

(ii) some $B \in \mathcal{B}$ satisties $s \le |B| \le v - s$.

Then $|\mathcal{B}| \ge \binom{v}{s}$.

Several people have observed that the concept of a t-design can be generalised as follows. Given a finite field $\mathbf{F}$, a t-design over $\mathbf{F}$ with parameters v, k, b_t is a collection of k-dimensional subspaces of a v-dimensional vector space over $\mathbf{F}$, called 'blocks', with the property that any t-dimensional subspace is contained in precisely b_t blocks; again I require $t \le k \le v - t$. Replacing 'design' with 'design over $\mathbf{F}$', $|B \cap B'|$ with $\dim(B \cap B')$, and the binomial coefficient $\binom{v}{s}$ with the function $\begin{bmatrix} v \\ s \end{bmatrix}_{\mathbf{F}}$ giving the number of s-dimensional subspaces of a v-dimensional vector space over $\mathbf{F}$, statements (1) and (2) remain valid, and their proofs require only trivial modifications. Similarly (3) can easily be converted into a valid statement:

(3') Let $\mathcal{B}$ be a collection of subspaces of a vector space X over $\mathbf{F}$, with $\dim(X) = v$, and s an integer, such that

(i) for $s \le i \le 2s$, the number of members of $\mathcal{B}$ containing a given i-dimensional subspace of X is a constant b_i, depending only on i;

(ii) some $B \in \mathcal{B}$ satisfies $s \le \dim(B) \le v - s$.

Then $|\mathcal{B}| \ge \begin{bmatrix} v \\ s \end{bmatrix}_{\mathbf{F}}$.

The proof I give below is essentially that of R. M. Wilson for the original statement (3). It was communicated to me by J. Doyen.

Suppose $|\mathbf{F}| = q$; let V and W be subspaces of the vector space X over $\mathbf{F}$, with $W \supseteq V$, $\dim(X) = a$, $\dim(W) = b$, $\dim(V) = c$. The number of subspaces U of X with $\dim(U) = d \ge c$, $U \cap W = V$, is

$$\frac{(q^a - q^b)(q^a - q^{b+1})\dots(q^a - q^{b+d-c-1})}{(q^d - q^c)(q^d - q^{c+1})\dots(q^d - q^{d-1})}.$$

We shall need two special cases of this. With $b = c$, the number of d-dimensional subspaces of X containing the b-dimensional subspace W is $\left[{a-b \atop d-b}\right]_{\mathbf{F}}$. With $b = a - 1$, $c = 0$, $d = 1$, the number of complements of a subspace of codimension 1 is q^{a-1}.

Lemma. With the hypotheses of (3'), if $s \leq i \leq i+j \leq 2s$, the number of members of $\mathcal{B}$ containing an i-dimensional subspace I and disjoint from a j-dimensional subspace J of X (with $I \cap J = \{0\}$) is a constant $b_{i,j}$ depending only on i and j. Furthermore, $b_{s,j} \neq 0$ for $0 \leq j \leq s$.

The proof is by induction on j; for $j = 0$ this is hypothesis (i), with $b_{i,0} = b_i$. Given I and J, with $\dim(I) = i$, $\dim(J) = j > 0$, and $i \geq s$, $i+j \leq 2s$, let J' be a subspace of J of codimension 1. Then $b_{i,j-1}$ members of $\mathcal{B}$ contain I and exclude J'; of these $b_{i+1,j-1}$ contain any complement for J' in J, and none contains more than one complement. So $b_{i,j} = b_{i,j-1} - q^{j-1} b_{i+1,j-1}$. The second statement is immediate from hypothesis (ii) (noting that there is an s-dimensional subspace disjoint from the given block B).

Proof of (3'). Let X_s denote the set of s-dimensional subspaces of X, and $\mathbf{Q}X_s$ a rational vector space with X_s as basis; $\dim(\mathbf{Q}X_s) = \left[{v \atop s}\right]_{\mathbf{F}}$. For $B \in \mathcal{B}$ define

$$\sigma_B = \sum \{S \mid S \in X_s,\ S \subseteq B\}.$$

It is enough to show that $\{\sigma_B \mid B \in \mathcal{B}\}$ spans $\mathbf{Q}X_s$, and so enough to show that, given $S_0 \in X_s$, $S_0 \in \langle \sigma_B \mid B \in \mathcal{B}\rangle_{\mathbf{Q}}$. For $0 \leq j \leq s$, put

$$\begin{aligned}\beta_j &= \sum \{\sigma_B \mid B \in \mathcal{B},\ \dim(B \cap S_0) = s - j\} \\ &= \sum_{\dim(B \cap S_0) = s-j} \ \sum_{S \subseteq B} S \\ &= \sum_{j \leq i \leq s} \ \sum_{\dim(S \cap S_0) = s-i} \left[{i \atop i-j}\right]_{\mathbf{F}} b_{s+i-j,j} S \qquad (*) \\ &= \sum_{j \leq i \leq s} \left[{i \atop i-j}\right]_{\mathbf{F}} b_{s+i-j,j}\, \varepsilon_i\end{aligned}$$

where

$$\varepsilon_i = \sum_{\dim(S \cap S_0) = s - i} S, \quad 0 \leq i \leq s.$$

This can be regarded as a system of linear equations for the ε's in terms of the β's. The coefficient matrix is triangular, with diagonal elements $b_{s,j} \neq 0$, and so is non-singular. Thus the equations have a solution. In particular,

$$S_0 = \varepsilon_0 \in \langle \beta_0, \ldots, \beta_s \rangle_{\mathbf{Q}} \subseteq \langle \sigma_B \mid B \in \mathcal{B} \rangle_{\mathbf{Q}} .$$

Comment on (*). Given S with $\dim(S) = s$, $\dim(S \cap S_0) = s - i$, we can choose in $\left[{s \quad - (s-i) \atop (s-j) - (s-i)}\right]_{\mathbf{F}}$ ways a subspace T of S_0 with $S \cap S_0 \subseteq T$, $\dim(T) = s - j$. Then we want to count blocks B with $\langle S, T \rangle \subseteq B$, $B \cap S_0 = T$. We can ensure the second condition by requiring B to be disjoint from some chosen complement of T in S_0. Note that we need the fact that $\langle S, T \rangle \cap S_0 = T$ to ensure that $\langle S, T \rangle$ is disjoint from the chosen complement.

We have proved a little more: if $\mathcal{B}$ satisfies the hypotheses of (3'), then the incidence matrix of s-dimensional subspaces against members of $\mathcal{B}$ (with incidence = inclusion) has rank $\left[{v \atop s}\right]_{\mathbf{F}}$. A similar remark applies to (3). This generalises a result of Kantor who showed it when $\mathcal{B}$ is the set of all k-dimensional subspaces (or k-subsets), with $s \leq k \leq v-s$.

We might try to generalise this result from the lattice of subsets of a set or subspaces of a vector space to wider classes of lattices with well-behaved rank functions (for example, geometric lattices). The results may generalise, but the proofs do not. We have already noticed a point in the above proof where we require modularity of the lattice; it will be obvious to the reader that something stronger than relative complementation has been freely used as well. (Of course, these properties characterise the lattices we have already considered.) Still, it might be worth looking at the lattice of subspaces of an affine geometry (noting that Kantor's result holds in this case), or at other lattices.

Unfortunately, I have no examples even of 2-designs over vector spaces except for the 'complete designs' whose blocks are all the k-dimensional subspaces. (1-designs are common: spreads, or subspaces of given type with respect to a symplectic form, are examples.) A 2-design over **F** gives rise to an ordinary 2-design on the set of 1-dimensional subspaces, when a block is identified with the set of 1-dimensional subspaces it contains. It is easy to prove that the case of equality in (1) with $s = 1$ over a finite field cannot occur. (Weakening the non-degeneracy condition, we might say that a symmetric design over a finite field must be a 'point-hyperplane design', that is, $k = v - 1$: these are trivial as designs over **F**, though highly non-trivial as ordinary symmetric designs.) I conjecture that it is provable that equality never occurs in (1) with $s \geq 1$ over any finite field. This is true for $s = 2$, though the proof I have is messy.

More general results have been obtained by Delsarte who extended these ideas, not to lattices, but to association schemes. His bound is given as the solution to a certain linear programming problem, though in some special cases (such as (1)) it can be converted into an explicit bound.

Merton College,
Oxford OX1 4JD

ON BALANCED ARRAYS*

D. V. CHOPRA

Summary

In this paper we derive some necessary conditions for the existence of balanced arrays of strength four, with two symbols and nine constraints.

1. Introduction and preliminaries

Balanced arrays are of much more recent origin as compared to orthogonal arrays. It will therefore be useful to define the former first.

Definition 1.1. A matrix T of size $(m \times N)$ with two elements (say, 0 and 1) is called a <u>balanced array</u> (B-array) with two symbols, m rows (constraints), N columns (runs or treatment-combinations) and strength t if for every submatrix $T_0(t \times N)$ of T and for every vector $\boldsymbol{\alpha}$ of size $(t \times 1)$ and weight i $(0 \le i \le t)$ (by the 'weight of a vector' we mean the number of nonzero elements in it), we have

$$(1.1) \quad \lambda(\boldsymbol{\alpha}, T_0) = \text{constant} = \mu_i^t \text{ (say)},$$

where $\lambda(\boldsymbol{\alpha}, T_0)$ is the number of columns of T_0 which are of weight i, and type $\boldsymbol{\alpha}$. The vector of nonnegative integers $\boldsymbol{\mu}' = (\mu_0^t, \mu_1^t, \ldots, \mu_t^t)$ is called the <u>index set</u> of the array.

Definition 1.2. A B-array is called an <u>orthogonal array</u> if $\mu_i^t = \mu$ (say) for each i. Clearly an orthogonal array is a special case of a balanced array. It is also obvious that N must be a multiple of 2^t. The positive integer $\mu = N 2^{-t}$ is known as the <u>index set</u> of the array.

* This work was partly supported by the Research Council of Wichita State University and the Department of Statistics at Colorado State University.

We may remark here, however, that the above definition of B-array with two symbols can be easily extended to that with s symbols.

Next, we discuss briefly the connection of B-arrays with other branches of the combinatorial theory of design of experiments.

Consider first the orthogonal arrays with $t = 2$. This class of arrays are equivalent to Hadamard matrices. On the other hand, the case $t = 2$ and $N = s^2$ corresponds to a set of $(m - 2)$ mutually orthogonal Latin squares of order s.

Next, let us consider B-arrays with $t = 2$. It can be shown in this case that the array is the incidence matrix of a balanced incomplete block design (BIBD) with possibly unequal block sizes. In the ordinary BIBD's the block sizes are taken to be all equal, and thus correspond to arrays with the additional restriction that each column be of equal weight.

B-arrays are quite useful in the theory of fractional factorial statistical designs. Indeed, B-arrays of strength $t = 2$, 3, and 4 are identical with balanced fractional factorial designs of resolutions 3, 4, and 5, respectively. We shall, however, not discuss here this aspect of balanced arrays and the interested reader may refer to Srivastava and Chopra (1971a, b).

It might be important to stress here that B-arrays provide a mathematically challenging field of research uniting various branches of the combinatorial theory of design of experiments and are very useful in practical problems arising in factorial experimentation. For some literature on B-arrays in general, the interested reader is referred to Srivastava (1972), Rafter (1971), and Srivastava and Chopra (1972).

Next, we present without proof, some results from Srivastava (1972), for later use. In what follows, we shall confine ourselves, in general, to B-arrays with $t = 4$ and $m = 9$, and the index set of such arrays is taken to be $\boldsymbol{\mu}' = (\mu_0, \mu_1, \mu_2, \mu_3, \mu_4)$. It is obvious that for such an array, we have

(1. 2) $N = \mu_0 + 4\mu_1 + 6\mu_2 + 4\mu_3 + \mu_4 = \mu'' + 4\mu' + 6\mu_2$, where $\mu' = \mu_1 + \mu_3$ and $\mu'' = \mu_0 + \mu_4$.

Theorem 1.1. Consider an array $T\ (9 \times N)$ of strength four. Let x_i denote the number of columns of T of weight i $(i=0, 1, 2, \ldots, 9)$ so that $\sum_{i=0}^{9} x_i = N$. Then the following single diophantine equations (SDE) must hold:

$$\text{(1.3)} \qquad \sum_{i=0}^{9} \binom{i}{j}\binom{9-i}{4-j} x_i = \binom{9}{4}\binom{4}{j} \mu_j, \quad j = 0, 1, 2, 3, 4.$$

Definition 1.3. An array T with two symbols 0 and 1, and m rows is called trim if $x_0 = x_m = 0$.

Remark. It is quite obvious that a 9-rowed B-array with N columns exists if and only if the corresponding trim B-array exists.

The SDE of (1.3), when written explicitly (when $x_0 = x_9 = 0$), become

$$\text{(1.4)} \quad \begin{array}{ll} \text{(a)} & 70x_1 + 35x_2 + 15x_3 + 5x_4 + x_5 = 126\mu_0 \\ \text{(b)} & 56x_1 + 70x_2 + 60x_3 + 40x_4 + 20x_5 + 6x_6 = 504\mu_1 \\ \text{(c)} & 21x_2 + 45x_3 + 60x_4 + 60x_5 + 45x_6 + 21x_7 = 756\mu_2 \\ \text{(d)} & 6x_3 + 20x_4 + 40x_5 + 60x_6 + 70x_7 + 56x_8 = 504\mu_3 \\ \text{(e)} & x_4 + 5x_5 + 15x_6 + 35x_7 + 70x_8 = 126\mu_4 \end{array}$$

We also recall the following from Srivastava and Chopra (1971a).

Theorem 1.2. Consider a B-array $T\ (9 \times N)$ with $t = 4$. Then we have

$$\text{(1.5)} \quad \begin{array}{ll} \text{(a)} & \mu_1 + \mu_3 \geq \mu_2 \\ \text{(b)} & 5\mu_2^2 \leq \mu_2(\mu_1 + \mu_3) + 7\mu_1\mu_3 \\ \text{(c)} & 72\mu' \leq 46\mu'' + 100\mu_2. \end{array}$$

Furthermore, we shall restrict ourselves to $\mu_2 \leq 6$. The reason for this is that $\mu_2 \leq 6$ corresponds to small values of N, a feature which makes a resolution V factorial design economic and thus usually very desirable.

2. Combinatorial analysis of trim arrays

In this section, we make certain investigations on trim B-arrays, which in turn are useful in obtaining other (nontrim) arrays.

In (1. 4), we make the following transformations

$$\begin{aligned} &u = x_2 + x_7, \quad v = x_3 + x_6, \quad w = x_4 + x_5, \quad s = x_1 + x_8 \\ &\mu' = \mu_1 + \mu_3, \quad \mu'' = \mu_0 + \mu_4 \quad \text{so that } N = u + v + w + s. \end{aligned} \tag{2.1}$$

From (1. 4a-e), we obtain

$$7u + 15v + 20w = 252\mu_2 \tag{2.2}$$

$$70s + 35u + 15v + 6w = 126\mu'' \tag{2.3}$$

$$28s + 35u + 33v + 30w = 252\mu'. \tag{2.4}$$

We obtain from (2. 2) and (2. 4):

$$3v + 7u + 8s = 72(\mu' - \tfrac{3}{2}\mu_2) \geq 0, \quad \text{i.e.} \quad \mu' \geq \tfrac{3}{2}\mu_2. \tag{2.5}$$

Next, (2. 3) and (2. 4) give us

$$23s + 10u + 3v = 9(5\mu'' - 2\mu') \geq 0, \quad \text{i.e.} \quad \mu'' \geq \tfrac{2}{5}\mu'. \tag{2.6}$$

It is quite obvious from (2. 5) and (2. 6) that

$$\mu'' \geq \tfrac{3}{5}\mu_2. \tag{2.7}$$

Furthermore, subtracting (2. 5) from (2. 6), we have

$$5s + u = 3[5\mu'' - 10\mu' + 12\mu_2] \geq 0, \quad \text{i.e.} \quad 10\mu' \leq 5\mu'' + 12\mu_2. \tag{2.8}$$

Also, (2. 5) and (2. 6) give us

$$11u + 45v = 36[50\mu' - 69\mu_2 - 10\mu''], \quad \text{i.e.} \quad 50\mu' \geq 69\mu_2 + 10\mu''. \tag{2.9}$$

From (2. 9), we observe that $9 | u$ so that

$$u = 9k \text{ where } k \text{ is a nonnegative integer.} \tag{2.10}$$

Using (2. 10) in (2. 8), we have

(2.11) $5s = 3[5\mu'' - 10\mu' + 12\mu_2 - 3k] \geq 0$, i.e. $0 \leq 3k \leq 5\mu'' + 12\mu_2 - 10\mu'$.

From (2.11), we have $3|s$, i.e.

(2.12) $s = 3k_1$ where k_1 is a nonnegative integer

(2.13) $5k_1 + 3k = 5\mu'' - 10\mu' + 12\mu_2$.

Next, we derive some simple but stringent conditions for the existence of such arrays.

Theorem 2.1. _If_ T _is a_ B-_array of size_ $(9 \times N)$, _then_

(2.14) $\max(\frac{63\mu_2}{5}, \frac{9\mu''}{5}) \leq N \leq 9\mu'$.

Proof. This follows immediately from (2.2-2.4).

Theorem 2.2. _Consider a_ B-_array_ T _of size_ $(9 \times N)$, _strength four, and with_ $\mu_2 \geq 5$. _Then_ $N \geq 66$.

Proof. From (2.5) and (2.7), we have $\mu' \geq 8$, $\mu'' \geq 3$, respectively. Take $\mu' = 8$, $\mu'' = 3$. This pair contradicts (2.8). Thus $\mu'' \geq 4$. Finally using (1.2), we have $N \geq 66$.

Theorem 2.3. _There exist no_ B-_arrays of size_ $(9 \times N)$, _with_ $\mu_2 = 4$ _and_ $N \leq 56$.

Proof. As in Theorem 2.2, we obtain $\mu' \geq 6$ and $\mu'' \geq 3$. Consider the arrays with $\mu' = 6$. From (2.14), we obtain $51 \leq N \leq 54$ and, therefore, $3 \leq \mu'' \leq 6$. That the arrays with $\mu' = 6$, $3 \leq \mu'' \leq 6$ are not possible is observed from the fact that there does not exist any 8-rowed array with $\mu_2 = 4$ and $N < 56$ (see Srivastava and Chopra, 1972). Next, we consider $\mu' = 7$. From (2.7) and (2.8), we have $\mu'' \geq 5$ and hence $N \geq 57$.

Theorem 2.4. _Consider a_ B-_array_ T $(9 \times N)$ _with_ $\mu_2 = 3$. _Then_ $N \geq 53$.

Proof. Here, as in above, we have $\mu' \geq 5$ and $\mu'' \geq 2$. That $(\mu', \mu'') = (5, 2)$ is not possible is obvious from (2.8). Therefore, we consider arrays with $\mu' = 5$ and $\mu'' \geq 3$. Here we have $41 \leq N \leq 45$. If these arrays exist, then deleting any one row would contain an 8-rowed subarray with $41 \leq N \leq 45$. This is contradicted by the fact that there does not exist any 8-rowed array with $\mu_2 = 3$ and $N \leq 59$ (see Chopra, 1973). The case of $\mu' = 6$ is rejected the same way. Therefore, we must have $\mu' \geq 7$ which, in turn, implies that $\mu'' \geq 7$. This proves the result.

Theorem 2.5. For a trim B-array T of size $(9 \times N)$ and $\mu_2 = 2$, we must have $N \geq 51$.

Proof. Here we observe $\mu' \geq 3$ and $\mu'' \geq 2$. For $3 \leq \mu' \leq 6$, we have $N \leq 54$. None of these arrays is possible can be seen from the fact that there does not exist any $(8 \times N)$ array with $\mu_2 = 2$ and $N < 56$ (see Chopra, 1973 and Chopra and Srivastava, 1972). Hence, $\mu' \geq 7$ and, therefore, $\mu'' \geq 11$. This proves the result.

Theorem 2.6. There exists no trim B-arrays of size $(9 \times N)$ with $\mu_2 = 1$ and $N < 36$.

Proof. First of all we observe (see Srivastava and Chopra, 1972) that for a B-array T $(8 \times N)$ with two symbols, $t = 4$ and $\mu_2 = 1$ we must have $N \geq 28$. From (2.14), we observe that $\mu' \geq 4$ and, therefore, $\mu'' \geq 6$. Take $\mu' = 4$, $\mu'' = 6$. From (2.13), we have $5k_1 + 3k = 2$, which is clearly not possible. Similarly the case $(\mu', \mu'') = (4, 7)$ is rejected. Thus $\mu'' \geq 8$. A similar argument would lead to the nonexistence of arrays with $8 \leq \mu'' \leq 13$. This establishes the result.

References

1. R. C. Bose and K. A. Bush. Orthogonal arrays of strength two and three. Ann. Math. Statist., 23 (1952), 508-24.
2. D. V. Chopra and J. N. Srivastava. Optimal balanced 2^7 fractional factorial designs of resolution V, with $N \leq 42$. To appear

in Ann. Inst. Statist. Math. (1971).

3. D. V. Chopra. Balanced optimal 2^8 fractional factorial designs of resolution V, $52 \leq N \leq 59$. To appear in the Proceedings of the International Symposium on Statistical Designs and Linear Models, held in Fort Collins, U. S. A. (1973).

4. D. V. Chopra and J. N. Srivastava. Optimal balanced 2^8 fractional factorial designs of resolution V, $37 \leq N \leq 51$. To appear in Sankhya (1973).

5. P. Dembowski. Finite Geometries. Springer-Verlag, New York (1968).

6. J. A. Rafter. Contributions to the theory and construction of partially balanced arrays. Ph. D. dissertation (under Professor E. Seiden). Michigan State University (1971).

7. C. R. Rao. Factorial arrangements derivable from combinatorial arrangements of arrays. Suppl. J. Roy. Statist. Soc., 9 (1947), 123-39.

8. C. R. Rao. Some combinatorial problems of arrays and applications to design of experiments. A Survey of Combinatorial Theory, edited by J. N. Srivastava et al., North-Holland Publishing Co., Amsterdam (1972).

9. E. Seiden and R. Zemach. On orthogonal arrays. Ann. Math. Statist., 37 (1966), 1355-70.

10. J. N. Srivastava. Optimal balanced 2^m fractional factorial designs. S. N. Roy Memorial Volume. University of North Carolina and Indian Statl. Inst., (1970), 227-41.

11. J. N. Srivastava. Some general existence conditions for balanced arrays of strength t and 2 symbols. J. Comb. Theory, 12, (1972), 198-206.

12. J. N. Srivastava and D. V. Chopra. On the characteristic roots of the information matrix of 2^m balanced factorial designs of resolution V, with applications. Ann. Math. Statist., 42 (1971a), 722-34.

13. J. N. Srivastava and D. V. Chopra. Balanced optimal 2^m fractional factorial designs of resolution V, $m \leq 6$. Technometrics 13 (1971b), 257-69.

14. J. N. Srivastava and D. V. Chopra. Balanced arrays and orthogonal arrays. A Survey of Combinatorial Theory, edited by J. N. Srivastava et al., North-Holland Publishing Co., Amsterdam (1972).

Wichita State University,
Kansas, U. S. A.

POSITIONS IN ROOM SQUARES

R. L. CONSTABLE

1. Introduction

A Room square of order $2n$, where n is a positive integer is an arrangement of $2n$ objects in a square array of side $2n - 1$ such that each of the $(2n - 1)^2$ cells of the array either is empty or contains exactly two distinct objects; each of the $2n$ objects appears exactly once in each row and column; and each (unordered) pair of objects occurs in exactly one cell. The objects are conventionally denoted by $\infty, 0, 1, \ldots, 2n-2$.

A cyclic Room square (CRS) is one in which the entry in cell $(i+1, j+1)$ is found by taking the entries in cell (i, j) and adding to each 1 modulo $2n - 1$ where $\infty + 1$ is defined to be ∞. The top row of the square has entry $(\infty, 0)$ in the left-hand or 0th position. The other occupied positions on the top row are denoted by p_i; $i = 1, 2, \ldots, n-1$.

A patterned Room square (PRS) is a CRS in which $x_i + y_i = 2n - 1$; $i = 1, 2, \ldots, n-1$.

Examples of a CRS and a PRS of order 8 are

(∞, 0)	-	-	(4, 6)	-	(2, 3)	(1, 5)
(2, 6)	(∞, 1)	-	-	(5, 0)	-	(3, 4)
(4, 5)	(3, 0)	(∞, 2)	-	-	(6, 1)	-
-	(5, 6)	(4, 1)	(∞, 3)	-	-	(0, 2)
(1, 3)	-	(6, 0)	(5, 2)	(∞, 4)	-	-
-	(2, 4)	-	(0, 1)	(6, 3)	(∞, 5)	-
-	-	(3, 5)	-	(1, 2)	(0, 4)	(∞, 6)

and

(∞, 0)	-	-	(2, 5)	-	(1, 6)	(3, 4)

A CRS of side 3 or 5 is impossible. Those of all other odd sides can be constructed, see [2], [3].

2. Positions

Theorem 1. If p_i, $i = 1, 2, \ldots, n-1$, are the positions of the occupied cells in the top row of a CRS, then $\Sigma p_i \equiv 0 (\text{mod } 2n - 1)$.

Proof. If the pair (x_i, y_i) occurs at p_i in the top row, then $(x_i - p_i + k_x, y_i - p_i + k_y)$ is the corresponding pair in the left hand column where

$$k_x, k_y = \begin{cases} 2n - 1 & \text{if the entry would otherwise be negative} \\ 0 & \text{if the entry is positive.} \end{cases}$$

The entries in the top row and left hand column, omitting the common entry $(\infty, 0)$, are two arrangements of the symbols $1, 2, \ldots, 2n-2$. Adding and equating row and column entries gives

$$\Sigma (x_i + y_i) = \Sigma (x_i + y_i - 2p_i) + w(2n - 1)$$

where w is the number of additions of $2n - 1$. Therefore, $\Sigma p_i = \frac{1}{2}w(2n - 1)$. Since w must be even $\Sigma p_i \equiv 0(\text{mod } 2n -1)$.

Theorem 2. $\Sigma p_i^2 \equiv \Sigma p_i(x_i + y_i)(\text{mod } 2n - 1)$.

Proof. A similar procedure to Theorem 1 is adopted for x_i^2, y_i^2. Hence $\Sigma(x_i^2 + y_i^2) = \Sigma(x_i^2 + y_i^2 + 2p_i^2 - 2p_i(x_i + y_i)) + 2t(2n - 1)$ for some positive t. So $\Sigma p_i^2 \equiv \Sigma p_i(x_i + y_i)$ (mod $2n - 1$).

Corollary 2. In a PRS, $\Sigma p_i^2 \equiv 0$ (mod $2n - 1$).

3. Reversible squares

Definition. Two CRS are said to be reversible if (x_i, y_i) occurs at p_i in one and at $2n - 1 - p_i$ in the other for $i = 1, 2, \ldots, n-1$.

All PRS are reversible. But unpatterned reversible squares (RRS) are few in number. For side 11 there are 2 pairs out of the 76 non-PRS e. g.

$(\infty, 0)$	$(3, 6)$	-	$(7, 9)$	$(1, 2)$	$(4, 8)$	-	-	-	$(5, 10)$	-
$(\infty, 0)$	-	$(5, 10)$	-	-	-	$(4, 8)$	$(1, 2)$	$(7, 9)$	-	$(3, 6)$

and their complements

$(\infty, 0)$	-	$(6, 1)$	-	-	-	$(7, 3)$	$(10, 9)$	$(4, 2)$	-	$(8, 5)$
$(\infty, 0)$	$(8, 5)$	-	$(4, 2)$	$(10, 9)$	$(7, 3)$	-	-	-	$(6, 1)$	-

Theorem 3. For a RRS, $\Sigma p_i^2 \equiv 0 \pmod{2n - 1}$.

Proof. $\Sigma p_i^2 \equiv \Sigma(2n - 1 - p_i)^2 \pmod{2n - 1}$. Hence $\Sigma p_i(x_i + y_i) \equiv \Sigma p_i^2 \equiv \Sigma(2n - 1 - p_i)^2 \equiv \Sigma(2n - 1 - p_i)(x_i + y_i)$ by theorem 2. Therefore, $2\Sigma p_i(x_i + y_i) \equiv \Sigma(2n - 1)(x_i + y_i) \equiv 0 \pmod{2n - 1}$. So $\Sigma p_i(x_i + y_i) \equiv 0 \pmod{2n - 1}$, and the result follows.

Theorem 4. A RRS with side $2n - 1$ exists if $2n - 1$ is a prime of form $4t + 3$ and, mod $2n - 1$, there is a primitive element g, where $1 + g^2$ is an even power of g. Furthermore the cell entry (g^{2r}, g^{2r+1}) will occupy position g^{2r+2} or g^{2r-1} if $1 + g$ is an even or odd power of g for $r = 0, 1, \ldots, n-2$.

The proof consists in showing that any difference $p_i - p_j$ is not congruent modulo $2n - 1$ to any difference between cell entries; see [1].

4. Skew squares

Definition. A CRS is said to be skew if the occupied positions are the 0th position along with one from each of the pairs $(i, 2n - 1 - i)$; $i = 1, \ldots, n-1$.

Theorem 5. There are no skew PRS with side size a multiple of 3.

Proof. For a skew CRS, p_i equals i or $2n - 1 - i$. But $i^2 \equiv (2n - 1 - i)^2 \pmod{2n - 1}$. Therefore, for a skew CRS, the squares of the occupied positions are an ordering of the squares of the integers $1, 2, \ldots, n-1$. Thus $\Sigma p_i^2 = \frac{1}{6}n(n - 1)(2n - 1)$. If $2n - 1 = 3s$, $\Sigma p_i^2 = \frac{1}{8}s(3s - 1)(3s + 1) \equiv s \pmod{3s}$, contrary to corollary 2. Hence, a PRS of side $3s$ cannot be skew.

References

1. R. L. Constable. Positions in Room squares. Utilitas Mathematica (to appear).

2. W. D. Wallis. On the existence of Room squares. Aequationes Math. (to appear).

3. W. D. Wallis. Solution of the Room square existence problem (to appear).

University of St. Andrews,
Fife, Gt. Britain

ANALOGUES OF HEAWOOD'S THEOREM

R. J. COOK

1. Introduction

One of the central problems which has stimulated the growth of graph theory is the 4-colour conjecture for planar maps. In one of the classic papers of graph theory Heawood [9] pointed out an error in Kempe's proof, resurrected enough to prove that planar graphs are 5-colourable and proved the following result for graphs of positive genus.

Theorem 1. Let G be a graph of positive genus γ. Then the chromatic number $\chi(G)$ of G satisfies

$$\chi(G) \leq [\tfrac{1}{2}(7 + \sqrt{\{1 + 48\gamma\}})],$$

where [x] denotes the integer part of x.

A close inspection of the proof reveals that it contains the following variation.

Theorem 2. Let G be a graph of positive genus γ. Then the minimum degree $\delta(G)$ of G satisfies

$$\delta(G) \leq [\tfrac{1}{2}(5 + \sqrt{\{1 + 48\gamma\}})].$$

For convenience we put

$$D(\gamma) = [\tfrac{1}{2}(5 + \sqrt{\{1 + 48\gamma\}})].$$

Most authors write $H(\gamma)$ for $1 + D(\gamma)$. However, we shall adopt this different notation because we believe that it is $D(\gamma)$ which is essential in the problem.

Theorem 1 is then deduced from Theorem 2 by showing that if $\chi(G) = n$ then G contains an n-critical subgraph G' and that G' has minimum degree $\delta(G') \geq n - 1$. Then

$$\chi(G) = n \leq 1 + D(\gamma') \leq 1 + D(\gamma).$$

2. Theme

The proof of Theorem 2 is very simple, it depends on three lemmas.

(i) Euler's formula: If G is a connected graph of genus γ with p points, q lines and r faces then

$$2 - 2\gamma = p - q + r. \tag{1}$$

(ii) A counting argument shows that if $\delta(G) \geq k$ then

$$2q \geq kp. \tag{2}$$

(iii) Since each face of G has at least 3 lines on its boundary

$$2q \geq 3r. \tag{3}$$

However, when Heawood proved Theorem 1 he believed that the inequality was sharp. This problem is far more difficult and was finally settled by Ringel and Youngs [18] in 1968, by showing that a suitable complete graph K_p may be embedded in an orientable surface of genus γ.

3. First variation

Euler's formula (1) holds for a graph embedded in an orientable surface of genus γ. If the graph G is embedded in a non-orientable surface of genus γ we have

$$2 - \gamma = p - q + r. \tag{4}$$

The corresponding bound for $\delta(G)$ is then, for $\gamma > 2$,

$$D^*(\gamma) = [\tfrac{1}{2}(5 + \sqrt{\{1 + 24\gamma\}})]. \tag{5}$$

The question of whether the corresponding bounds are sharp was settled by Ringel [17] in 1960, when he embedded suitable complete graphs K_p

in non-orientable surfaces. Thus all theorems of Heawood type have non-orientable analogues.

4. Second variation

Heawood's theorem was deduced from Theorem 2 by bounding the chromatic number of a critical subgraph by its minimum degree. Many other graph-theoretic parameters may be treated similarly.

(i) Point-arboricity $\rho(G)$ of a graph G is the minimum number of subsets in any partition of the vertex set of G so that each subset induces an acyclic subgraph. Chartrand and Kronk [2] have shown that if G is n-critical for point-arboricity (i. e. if $\rho(G) = n$ and G is a minimal graph with this property) and $n \geq 2$ then $\delta(G) \geq 2(n - 1)$. The corresponding analogue of Heawood's theorem is due to Kronk [10].

(ii) Point partition numbers $\rho_k(G)$ were introduced by Lick and White [14] as generalizations of point-arboricity and chromatic number. They showed that if G is n-critical for ρ_k then

$$\delta(G) \geq (k + 1)(n - 1)$$

and in [15] they deduced the corresponding analogues of Heawood's theorem.

(iii) The point (line) connectivity $\kappa(G)$ $(\lambda(G))$ of a graph G is the minimum number of points (lines) whose removal disconnects G. The point-connectivity, line-connectivity and the minimum degree of G are related by the inequalities

$$\kappa(G) \leq \lambda(G) \leq \delta(G), \qquad (6)$$

see Whitney [11]. The corresponding analogue of Heawood's theorem is proved in [7]. It may be worthwhile remarking that Geller and Harary [8] have obtained inequalities for digraphs corresponding to (6).

Clearly, similar results may be obtained for any parameter which can be bounded above by the minimum degree. For example, Chvátal [3] introduced the toughness $\tau(G)$ of a graph G as follows:

Let t be a real number such that the implication

$$k(G - S) > 1 \Rightarrow |S| \geq t\,k(G - S) \qquad (7)$$

holds for each set S of points of G, where k(H) is the number of components of the graph H. Then G is said to be t-tough. If G is not complete then there is a largest t such that G is t-tough, and this number is called the toughness of G. Pippert [16] has shown that if G is not a complete graph then $\tau(G) \leq \frac{1}{2}\delta(G)$. Thus if G is a graph of positive genus γ and G is not a complete graph then

$$\tau(G) \leq \tfrac{1}{2}D(\gamma). \tag{8}$$

5. Third variation

The girth g(G) of a graph G is the length of the shortest cycle, if any, in G; if G is acyclic then g(G) is not defined. If G has girth g then every face of G has at least g lines on its boundary and so

$$2q \geq gr. \tag{9}$$

Various aspects of this case have been studied in [4]-[7] and [11]-[13]. If G is acyclic then $\delta(G) \leq 1$, so we may suppose that G contains a cycle. If G is triangle-free then $g(G) \geq 4$ and Euler's equation (1) becomes

$$2 - 2\gamma \leq p - \tfrac{1}{2}q$$

so that if $\delta(G) \geq k$,

$$8(\gamma - 1) + 4p \geq 2q \geq kp.$$

Turan [19] has shown that $q \leq p^2/4$ for triangle-free graphs so

$$\tfrac{1}{2}p^2 \geq 2q \geq kp$$

and so $p \geq 2k$. Then if $k \geq 4$

$$8(\gamma - 1) \geq (k - 4)p \geq 2k(k - 4),$$

i. e.

$$k^2 - 4k + 4(\gamma - 1) \leq 0.$$

Thus

$$k \leq 2 + [2\sqrt{\gamma}]. \qquad (10)$$

The condition $k \geq 4$ becomes $\gamma > 0$. For planar graphs we have $k \leq 3$. For $\gamma > 0$ the bound in (10) is sharp, a suitable example being the complete bipartite graphs $K_{n,n}$; moreover, they also show that the corresponding bounds for connectivity are sharp.

For general girth g the same methods give

$$p \leq 2qk^{-1} \leq 4g(\gamma - 1)/(gk - 2g - 2k), \qquad (11)$$

provided that $gk - 2g - 2k > 0$, and this is then combined with a lower bound for p.

Let G be a graph of girth g with minimum degree k and maximum degree K. Then $p \geq c(k, K, g)$ where

$$c(k, K, g) = 1 + K + K(k-1) + \ldots + K(k-1)^{\frac{1}{2}(g-3)} \quad \text{if } g \text{ is odd,}$$

and

$$= 2 + (K + k - 2)(1 + (k-1) + (k-1)^2 + \ldots + (k-1)^{\frac{1}{2}(g-4)}) \quad \text{if } g \text{ is even.}$$

This is proved in essentially the same way as the similar results for regular graphs which may be found in Biggs [1, §4. 2] and Tutte [20, §8. 3].

For any fixed values of k and γ these inequalities are not both soluble if g is sufficiently large, and it is thus possible to deduce a bound for the minimum degree $\delta(G)$. The corresponding results for the various other parameters are then deduced in the usual way.

References

1. Norman Biggs. Finite groups of automorphisms. London Math. Soc. Lecture Notes, 6 (1971) (Cambridge University Press).

2. Gary Chartrand and Hudson V. Kronk. The point-arboricity of planar graphs. J. London Math. Soc., 44 (1969), 612-6.

3. V. Chvátal. Tough graphs and Hamiltonian circuits. J. Discrete Math., (to appear).

4. R. J. Cook. Chromatic number and girth. (To appear.)

5. R. J. Cook. Point-arboricity and girth. J. London Math. Soc., (to appear).

6. R. J. Cook. Point partition numbers and girth. (To appear.)

7. R. J. Cook. Heawood's theorem and connectivity. Mathematika, (to appear).

8. D. Geller and F. Harary. Connectivity in digraphs. Recent trends in graph theory, Lecture Notes in Mathematics, 186 (1971), 105-15 (Springer-Verlag).

9. P. J. Heawood. Map colour theorem. Quart. J. Math., 24 (1890), 332-8.

10. Hudson V. Kronk. An analogue to the Heawood map-colouring problem. J. London Math. Soc., (2), 1 (1969), 750-2.

11. Hudson V. Kronk. The chromatic number of triangle-free graphs. Graph theory and applications, Lecture Notes in Mathematics, 303 (1972), 179-81 (Springer-Verlag).

12. Hudson V. Kronk and John Mitchem. Critical point-arboritic graphs. (To appear.)

13. Hudson V. Kronk and Arthur T. White. A 4-color theorem for toroidal graphs. Proc. Amer. Math. Soc., 34 (1972), 83-6.

14. D. R. Lick and A. T. White. k-degenerate graphs. Canadian J. Math., 22 (1970), 1082-96.

15. D. R. Lick and A. T. White. The point partition numbers of closed 2-manifolds. J. London Math. Soc., (2), 4 (1972), 577-83.

16. R. E. Pippert. On the toughness of a graph. Graph theory and applications, Lecture Notes in Mathematics 303 (1972), 225-33 (Springer-Verlag).

17. G. Ringel. Färbungsprobleme auf Flachen und Graphen. Deutscher Verlag der Wissenschaften, Berlin, 1962.

18. G. Ringel and J. W. T. Youngs. Solution of the Heawood map-coloring problem. Proc. Nat. Acad. Sci. USA, 60 (1968), 438-45.

19. P. Turan. Eine Extremalaufgabe aus der Graphentheorie. Mat. Fiz. Lapok, 48 (1941), 436-52.

20. W. T. Tutte. The connectivity of graphs. (University of Toronto Press, Toronto, 1966.)

21. H. Whitney. Congruent graphs and the connectivity of graphs. Amer. J. Math., 54 (1932), 150-68.

University College,
Cardiff, Wales

CUT-SET LATTICES OF GRAPHS

F. ESCALANTE

Let G be a connected, undirected, simple graph. Let V(G) and E(G) denote the vertex and edge sets respectively. Consider two different vertices a, b of G. A subset T of the vertex (edge) set is called an a, b-vertex (edge) cut if T separates a and b but no proper subset T' of T does, i.e. if in G - T the vertices a and b belong to different connected components but in G - T' there is always an a, b-path. Cuts with the smallest cardinality are called minimal.

Let now S, T be two a, b-vertex (edge) cuts and order them in such a way that $S \leq T$ if and only if no a, b-path meets T 'before' S. Then it can be shown that $\leq$ is a partial order; indeed we have (for proofs see for example [1]):

Theorem 1. Let Γ_1 and Γ_2 be the sets of all a, b-vertex cuts and a, b-edge cuts, respectively, of a graph G with respect to two different, fixed vertices a, b of G. Then $(\Gamma_1, \leq)$ and $(\Gamma_2, \leq)$ constitute complete lattices.

Theorem 2. Let Δ_1 and Δ_2 be the sets of all minimal a, b-vertex cuts and minimal a, b-edge cuts, respectively, of a graph G with respect to two different, fixed vertices a, b of G. If the cardinality of the cuts is finite, then $(\Delta_1, \leq)$ and $(\Delta_2, \leq)$ constitute distributive lattices.

Denote by $V_i(G; a, b)$ the lattice $(\Gamma_i, \leq)$ and by $U_i(G; a, b)$ the distributive lattice $(\Delta_i, \leq)$, $i = 1, 2$. Call a complete lattice L (a finite distributive lattice L) V_i-representable (U_i-representable) if there exists a simple graph G and $a, b \in V(G)$ such that $V_i(G; a, b) \cong L$ $(U_i(G; a, b) \cong L)$.

In [1] we proved:

Theorem 3. Every complete lattice is V_1-representable.

Theorem 4. Every finite distributive lattice is U_1-representable.

And in [2] we showed:

Theorem 5. Every finite distributive lattice is U_2-representable.

Recently I proved that in every $V_2(G; a, b)$ the chains between any two comparable elements have the same length, i. e. that $V_2(G; a, b)$ satisfies the Jordan's chain condition. This fact implies that not every complete lattice is V_2-representable and points out an algebraic difference between edge- and point-connectivity, which is surprising considering Theorems 4 and 5. The proof of this and related facts will be published elsewhere.

References

1. F. Escalante. Schnittverbände in Graphen. Abh. Math. Sem. Hamburg, 38 (1972), 199-220.
2. F. Escalante and T. Gallai. Note über Kantenschnittverbände in Graphen. Submitted to the Acta Math. Acad. Sci. Hung.
3. D. König. Theorie der endlichen und unendlichen Graphen. Leipzig, 1936.
4. R. Halin. Über trennende Eckenmengen in Graphen und den Mengerschen Satz. Math. Ann., 157 (1964), 34-41.
5. L. Lovasz. A remark on Menger's theorem. Acta Math. Acad. Sci. Hung., 21 (1970), 365-8.
6. J. S. Pym. A lattice of separating sets in a graph, in Combinatorial Mathematics and its Applications, Academic Press, 1971.

CIMAS,
National University of Mexico

ON THE CHROMATIC INDEX OF A GRAPH, II

STANLEY FIORINI and ROBIN J. WILSON

This article is a sequel to our earlier paper [5] in 'Cahiers du Centre d'Etudes de Recherche Opérationnelle'. We propose to extend and fill in the details of the proofs of the results announced in the former paper. As in that paper the object is twofold:

(A) to present a survey of old and new results in the classification problem arising from Vizing's theorem;

(B) to study some aspects of the structure of critical graphs which are very useful in this classification.

Notation. Throughout this paper we shall use the following notation. G denotes a simple graph whose vertex-set is V(G) and edge-set E(G); n and m denote respectively the cardinalities of these sets. $\rho(v)$ denotes the valency of a vertex v, and $\rho = \max \rho(v)$. $\sum_{v \in V(G)} \rho - \rho(v)$ is called the total deficiency. α denotes the edge-independence number, i. e. the largest number of edges no two of which have a vertex in common. K_n is the complete graph on n vertices, C_n the circuit graph on n vertices, and $K_{m,n}$ the complete bipartite graph on m and n vertices. $\hat{G}$ will denote the line-graph of the graph G. // denotes the end or absence of a proof.

For a more complete list of definitions one is referred to [17].

A The classification problem

1. Vizing's theorem

We define the chromatic index $\chi_e(G)$ of a graph G to be the least number of colours needed to colour the edges of G in such a way that any two adjacent edges are assigned different colours.

In 1964, Vizing [13], obtained very sharp bounds for the chromatic index of a graph:

Theorem 1. (Vizing) If G is a simple graph, then

$$\rho \le \chi_e(G) \le \rho + 1. \quad //$$

In view of this theorem we can classify graphs into either of two classes. We say that G is of class one if $\chi_e(G) = \rho$, and that G is of class two if $\chi_e(G) = \rho + 1$. We denote this by $G \in C^1$ or $G \in C^2$ (or by $G \in C^1_\rho$ or $G \in C^2_\rho$ if we want to emphasize the fact that $\max_{v \in V(G)} \rho(v) = \rho$). The problem of deciding which graphs belong to which class remains very much an unsolved problem. It suffices to point out that its complete solution would immediately settle the four-colour conjecture. This follows from the fact that the four-colour conjecture is true if and only if every cubic, planar, bridgeless graph is in C^1.

2. Some sufficient conditions

We give here some sufficient conditions for a graph to be in C^1 or C^2.

Theorem 2. (i) (Vizing [15]) If $\max_{G' \subseteq G} \min_{v \in V(G')} \rho(v) \le \frac{1}{2}.\rho$, then $G \in C^1_\rho$. //

(ii) (Welsh [7] p. 149) If G is not an odd circuit and if all the circuits of G have the same parity, then $G \in C^1$. //

Corollary 1. (i) If λ_G denotes the largest eigenvalue of the adjacency matrix of G, and $\lambda_G \le \frac{1}{2}.\rho$, then $G \in C^1$.

(ii) Bipartite graphs are in C^1.

(iii) (Vizing) Planar graphs with $\rho \ge 10$ are in C^1.

Remark. In [14], (iii) has been strengthened as follows: Planar graphs with $\rho \ge 8$ are in C^1.

Proof. (i) Let $\tilde{G}$ be the induced subgraph of G for which

$$\min_{v \in V(\tilde{G})} \rho(v) = \max_{G' \subseteq G} \min_{v \in V(G')} \rho(v).$$

Then $\lambda_G \ge \lambda_{\tilde{G}} \ge \min_{v \in V(\tilde{G})} \rho(v) = \max_{G' \subseteq G} \min_{v \in V(G')} \rho(v)$. The result then follows by Theorem 2 (i). //

(ii) This follows from Theorem 2 (ii). //

(iii) This can be proved using Theorem 2 (i); see [15] for further details. //

Theorem 3. (i) (Beineke and Wilson [1]) If G satisfies $n = 2k + 1$ and if the total deficiency is less than ρ, then $G \in C^2$. //

(ii) If $m/\alpha > \rho$, then $G \in C^2$.

(iii) Let G be regular and contain a set of t edges whose removal gives an odd, non-trivial component; if $t < \rho$, then $G \in C^2$.

Proof. (ii) Let $G \in C^1$. Then there exists a partition of $E(G)$ into ρ colour-classes $\{C_j\}_{j=1}^{\rho}$. Now $\alpha \geq |C_j|$. So

$$\sum_{j=1}^{\rho} \alpha \geq \sum_{j=1}^{\rho} |C_j| = m.$$

This implies that $m \leq \alpha\rho$. //

(iii) Let E be the separating set of t edges and let $G \backslash E = H_1 \cup H_2$; $H_1 \cap H_2 = \emptyset$; $|V(H_1)|$ be odd. Then H_1 has total deficiency less than ρ. So, by (i) $H_1 \in C_{\rho}^2$, which implies that $G \in C_{\rho}^2$. //

Corollary 2. (i) Regular graphs of odd order (e.g. K_{2k+1}, C_{2k+1}) are in C^2.

(ii) Regular graphs with a cut-vertex are in C^2.

(iii) Graphs of order $2k + 1$ and with $m > k.\rho$ are in C^2.

(iv) If G is obtained from a regular graph of even order by the insertion of a new vertex into any one of its edges, then G is in C^2.

Proof. (i), (iii) and (iv) follow from Theorem 3 (i); (ii) follows from Theorem 3 (iii). //

Counter-examples can be given for each of these conditions to show that they are only sufficient and not necessary. By an ad hoc construction, Berge [2], among others, shows that $K_{2k} \in C^1$. Also, a case-by-case analysis of all connected graphs of order not exceeding six enabled Beineke and Wilson to show that of a total of 143 such graphs only 8 are in C^2 (see [1]).

3. Constructions of C^2-graphs

As soon as we have a supply of C^2-graphs, we can combine them together to give other C^2-graphs. The following are some examples of how this can be achieved.

(i) If G is any graph in C^2_ρ, then any supergraph whose maximal valency does not exceed ρ is also in C^2_ρ. Erdős and Kelley [6], have given necessary and sufficient conditions on the number of vertices, maximal and total deficiency which yield such a regular supergraph of minimal order.

(ii) Any Hajos union of two C^2_ρ-graphs G_1, G_2 (denoted by $G_1 \cup_H G_2$ and defined as follows) gives another C^2_ρ-graph: Pick $v_i \in V(G_i)$ $(i = 1, 2)$ such that $\rho(v_1) + \rho(v_2) \leq \rho + 2$, and identify v_1 with v_2. Remove any edge (v_i, w_i) from $E(G_i)$ $(i = 1, 2)$, and join the vertices w_1 and w_2 (see the diagram). This construction was first

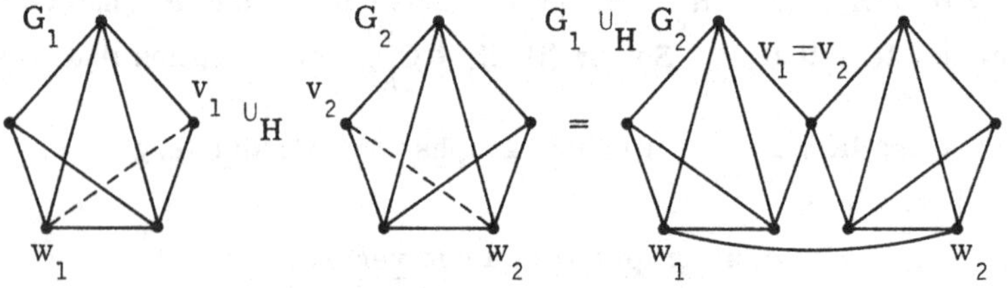

defined for vertex-colourings by Hajos and has recently been shown to hold also for edge-colourings by Jakobsen [8].

(iii) In [10], Meredith, constructs C^2_ρ-graphs which are regular of even order, ρ edge-connected and non-hamiltonian.

B Critical graphs

If G is a C^2_ρ-graph with the property that $G \setminus \{e\} \in C^1_\rho$ for each edge e, then G is said to be critical, and we write $G \in C^*_\rho$ or $G \in C^*$.

Just as in problems relating to the colouring of vertices of a graph, critical graphs play an important rôle in the theory of edge-colourings. This is to be expected, since if we define criticality as above, then any C^2_ρ-graph contains a C^*_k-subgraph for any integer k satisfying $2 \le k \le \rho$.

The usefulness of critical graphs is also evident from the fact that they have much more structure than arbitrary graphs. So by restricting ourselves to critical graphs we lose nothing and often gain a lot.

In our study of properties of C*-graphs we have taken as a model analogous results for critical graphs in vertex colourings. In particular, we propose to investigate questions of existence, connectivity and bounds on the number of edges of C*-graphs. In the following we make free use of the following easily proved result:

Theorem 4. Let G be a graph and $\hat{G}$ its line-graph; then

(i) $\chi_e(G) = \rho$ if and only if $\chi_v(\hat{G}) = \rho$;

(ii) $G \in C^*_\rho$ if and only if $\hat{G}$ is vertex-critical and $(\rho + 1)$-colourable. //

Unfortunately, not all results about vertex-colourings can be translated via the line-graph into results about edge-colourings.

1. Elementary properties of C*-graphs

We now prove some elementary properties of C*-graphs.

Theorem 5. If $G \in C^*$ then:

(i) G has no cut-vertex (and a fortiori is bridgeless);

(ii) $G \in C^*_2$ if and only if G is an odd circuit;

(iii) If $G \in C^*_\rho$ $(\rho \ge 3)$, then G cannot be regular;

(iv) If I is an arbitrary set of independent edges of G, then

$$\chi_e(G\backslash I) = \chi_e(G) - 1.$$

(v) G is not uniquely colourable, provided $G \ne K_3$.

Proof. (i) G has a cut-vertex v of degree q

$\Rightarrow \hat{G} = H_1 \cup H_2$, $H_1 \cap H_2 = K_q$

$\Rightarrow \hat{G}$ is not vertex-critical (Dirac [4])

$\Rightarrow G \notin C^*$. //

(ii) G is an odd circuit $\Longleftrightarrow$ $\hat{G}$ is an odd circuit

$\Longleftrightarrow$ $\hat{G}$ is vertex-critical and 3-colourable (Dirac [3])

$\Longleftrightarrow$ $G \in C^*_2$. //

(iii) Case 1. $|V(G)| = 2k + 1$, $G \in C^2_\rho$ $(\rho \geq 3)$, G regular

$\Rightarrow G\backslash\{e\}$ has total deficiency $< \rho$, where e is an edge of G

$\Rightarrow G\backslash\{e\} \in C^2_\rho$ (by Theorem 3 (i))

$\Rightarrow G \notin C^*_\rho$.

Case 2. $|V(G)| = 2k$, $G \in C^*_\rho$ $(\rho \geq 2)$, G regular

$\Rightarrow \tilde{G} = G\backslash\{(v, w)\} \in C^1_\rho$, where (v, w) is an edge of G

$\Rightarrow |\theta_v \cap \theta_w| = \rho - 2$ (Berge [2] Ch. 12), where θ_v is the set of colours of those edges incident with v

$\Rightarrow \tilde{G}$ has $(\rho - 2)$ perfect matchings $\{E_i\}_{i=1}^{\rho-2}$

$\Rightarrow \tilde{G}\backslash\{\bigcup_{i=1}^{\rho-2} E_i\}$ is regular of degree 2 (except for the vertices v and w which have degree 1) and is two-colourable

$\Rightarrow \tilde{G}\backslash\{\bigcup_{i=1}^{\rho-2} E_i\}$ is the disjoint union of even circuits and of a chain P, say.

If P has an even number of edges (and hence an odd number of vertices), we have a contradiction to the fact that $|V(G)| = 2k$; on the other hand, if P has an odd number of edges (and an even number of vertices), then we can introduce (v, w) to colour $G\backslash\{\bigcup_{i=1}^{\rho-2} E_i\}$ with two colours, which induces a ρ-coloration of G. This contradicts the fact that G is in C^2_ρ. //

(v) If $G \in C^*$, then there exists a coloration in which one colour-class consists precisely of any one pre-assigned edge. Hence, the colorations associated with the various edges induce different partitions of $E(G)$, provided not all edges are mutually adjacent, in which case each edge forms a colour-class by itself. This can happen in either of two ways: $G = K_3$, which we are excluding, or $G = K_{1,t}$, which is not in C^*. //

2. Bounds on $|E(G)|$

In [14], Vizing proved the following important results:

Theorem 6. *If* $G \in C^*_\rho$, $v, w \in V(G)$, $(v, w) \in E(G)$, $\rho(v) = k$, *then* w *is adjacent to at least* $(\rho - k + 1)$ *other vertices of degree* ρ. //

Corollary 3. (i) $|E(G)| \geq (3\rho^2 + 6\rho - 1)/8$;

(ii) G *has at least three vertices of degree* ρ;

(iii) *each vertex is adjacent to at least two vertices of degree* ρ. //

The bound in (i) is quite good if ρ is not too 'small' when compared with $(n - 1)$, but is poor for smaller values of ρ since it does not take the number of vertices into account. This is illustrated in graphs like:

$\bigcup_{i=1}^{k}{}_H\, G_i =$

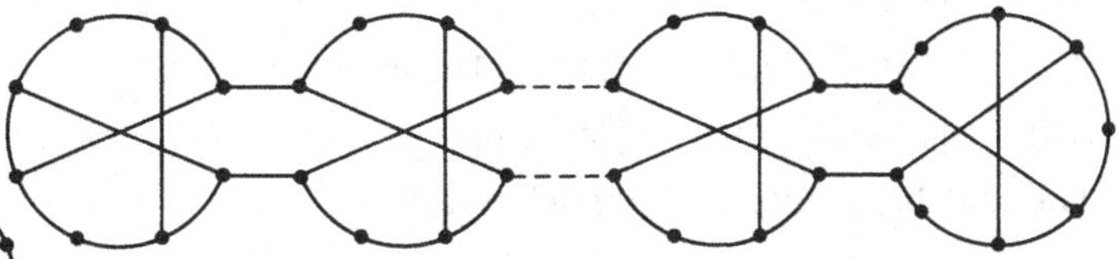

where $G_i =$

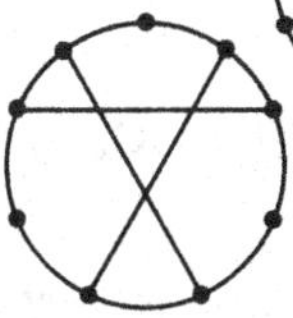

Later we shall present a result to complement the above result.

First we prove the following proposition:

Theorem 7. *If* $G \in C^*_\rho$ *and* $(v, w) \in E(G)$, *then*

$\rho(v) + \rho(w) \geq \rho + 2$.

Proof. $G \in C^*_\rho \Rightarrow \hat{G}$ is vertex-critical

$\Rightarrow \rho(x) \geq \rho$ for all vertices x (Ore [11])

$\Rightarrow \rho(\theta(e)) \geq \rho$ where θ is the one-one correspondence between $E(G)$ and $V(\hat{G})$

$\Rightarrow \rho(v) + \rho(w) - 2 \geq \rho$ for each $(v, w) \in E(G)$. //

In fact Berge [2] tells exactly by how much $\rho(v) + \rho(w)$ exceeds $\rho + 2$ in the following theorem:

Theorem 8. If θ_v denotes the set of colours of edges incident with vertex v in some edge-coloration of a C^*_ρ-graph G, then, for each $(v, w) \in E(G)$ we have:

(i) $|\theta_v \cap \theta_w| = (\rho(v) + \rho(w)) - (\rho + 2);$

(ii) $|\theta_v \cup \theta_w| = \rho;$

(iii) $|\theta_v \backslash \theta_w| = \rho + 1 - \rho(w);$

(iv) $|\theta_w \backslash \theta_v| = \rho + 1 - \rho(v).$ //

We have the following bounds on $|E(G)|$ if $G \in C^*_\rho$.

Theorem 9. Let $G \in C^*_\rho$ and let S denote $\sum_{i=1}^{n} (\rho(v_i))^2$ and T denote $\frac{\rho}{(\rho + 1)}$, then we have:

$$(T^2 + ST)^{\frac{1}{2}} - T \le |E(G)| \le (S/(\rho + 2) - \tfrac{1}{2}(\rho - 2)).$$

Proof. This follows by straightforward manipulation of:

(i) $|E(\hat{G})| + |E(G)| = \frac{1}{2}S;$

(ii) $|E(\hat{G})| \le m^2(\rho - 1)/2\rho$ (Turan [12]);

(iii) $|E(\hat{G})| \ge \frac{1}{2}m\rho + \frac{1}{4}(\rho^2 - 4).$ //

The last inequality is obtained as follows:

Let ϕ_ρ denote the number of vertices of degree ρ in G, and let σ denote the minimum vertex-degree of G. Then, $\min_{v \in V(\hat{G})} \rho(v) \ge \max\{\rho, 2\sigma-2\}$, since $\hat{G}$ is a line-graph which is vertex-critical. Also, by Theorem 6,

$$\phi_\rho \ge (\rho - \sigma + 1) + 1 = (\rho - \sigma + 2).$$

So,

$$\begin{aligned} 2.\ |E(\hat{G})| &\ge \phi_\rho(2\rho - 2) + (m - \phi_\rho).\max\{\rho, (2\sigma - 2)\} \\ &\ge (\rho - \sigma + 2).(2\rho - 2 - \max\{\rho, (2\sigma-2)\}) + m.\max\{\rho, (2\sigma-2)\} \\ &\ge m\rho + \tfrac{1}{2}(\rho^2 - 4). \quad // \end{aligned}$$

This result is quite good for testing particular graphs, but bounds which involve just n and ρ are desirable. To this end we have:

Lemma 1. If G is a C^*_ρ-graph, then $\sum_{j=2}^{k} \phi_j \le \frac{1}{2}(k - 1)\phi_\rho$ for any k satisfying $2 \le k \le (\rho - 1)$.

Proof. Let $\phi_\rho(i_1, i_2, i_3, \ldots, i_k)$ denote the number of vertices of degree ρ adjacent to precisely i_1 vertices of degree 1, i_2 vertices of degree 2, ..., and i_k vertices of degree k. Then by Theorem 6 and Corollary 3 (iii),

$$2.\phi_j \le \sum_{(k)} i_j.\phi_\rho(i_1, i_2, \ldots, i_j, \ldots, i_k),$$

where the summation is taken over all k-tuples whose entries form all partitions of integers h such that

$$h = \sum_{t=1}^{k} i_t \le \rho - (\rho - q + 1) = (q - 1).$$

Here q denotes the smallest index of all non-zero elements of the k-vector. So,

$$\sum_{j=1}^{k} 2.\phi_j \le \sum_{j=1}^{k} \sum_{(k)} i_j.\phi_\rho(i_1, \ldots, i_j, \ldots, i_k)$$

$$= \sum_{(k)} \sum_{j=1}^{k} i_j.\phi_\rho(i_1, \ldots, i_j, \ldots, i_k)$$

$$\le \sum_{(k)} (k - 1).\phi_\rho(i_1, \ldots, i_j, \ldots, i_k)$$

$$\le (k - 1).\phi_\rho. \quad //$$

From this we can deduce the following corollary:

Corollary 4. <u>If</u> $G \in C^*_\rho$, <u>then</u> $\phi_\rho \ge 2n/\rho$.

Proof. Put $k = (\rho - 1)$ in Lemma 1. //

We can now give a lower bound for $|E(G)|$ in terms of n and ρ as follows:

The following sum represents a lower bound for $2.|E(G)|$:

$$\rho.\phi_\rho + (2.\tfrac{1}{2}\phi_\rho + 3.\tfrac{1}{2}\phi_\rho + \ldots + h.\tfrac{1}{2}\phi_\rho) + (n-(h-1).\tfrac{1}{2}\phi_\rho - \phi_\rho).(h+1),$$

provided (i) $2 \le h \le (\rho - 1)$, (ii) $n - (h-1).\frac{1}{2}\phi_\rho - \phi_\rho \ge 0$, i.e. $2n \ge \phi_\rho.(h + 1)$. With these provisos:

$$2m \ge n.(h + 1) + \phi_\rho.(\rho - \tfrac{1}{4}(h^2 + 3h + 4)).$$

Now let $h = (2\rho)^{\frac{1}{2}} - 1$, $\phi_\rho \le 2n/(2\rho)^{\frac{1}{2}}$ and $\rho \ge 5$. Then (i) and (ii) are satisfied, and so:

$$m \ge n((\rho/2)^{\frac{1}{2}} + \frac{2\rho - (2\rho)^{\frac{1}{2}} - 2}{4\rho}) .$$

Now let $\phi_\rho \ge 2n/(2\rho)^{\frac{1}{2}}$. Then

$$2m \ge \rho . 2n/(2\rho)^{\frac{1}{2}} + 2(n - 2n/(2\rho)^{\frac{1}{2}})$$
$$\Rightarrow \quad m \ge n((\rho/2)^{\frac{1}{2}} + 1 - (2/\rho)^{\frac{1}{2}})$$
$$\ge n((\rho/2)^{\frac{1}{2}} + \frac{2\rho - (2\rho)^{\frac{1}{2}} - 2}{4\rho}).$$

Also, if $\rho = 2$, $m = n$;

if $\rho = 3$, $\phi_3 \ge 2n/3$, which implies that $m \ge 4n/3$;

if $\rho = 4$, $\phi_4 \ge n/2$, $\phi_3 \le \phi_4$, $\phi_2 \le \frac{1}{2} . \phi_4$, which implies that $m \ge 13n/8$.

So in all cases we have:

$$m \ge n. ((\rho/2)^{\frac{1}{2}} + \frac{2\rho - (2\rho)^{\frac{1}{2}} - 2}{4\rho}) .$$

This and Vizing's result, together imply:

Theorem 10. <u>If</u> $G \in C^*_\rho$, <u>then</u>

$$|E(G)| \ge \max\{n. ((\rho/2)^{\frac{1}{2}} + \frac{2\rho - (2\rho)^{\frac{1}{2}} - 2}{4\rho}), \frac{(3\rho^2 + 6\rho - 1)}{8}\} .$$

In fact Vizing [16] conjectures the following stronger bound:

$$|E(G)| \ge \frac{n(\rho - 1) + 3}{2} .$$

This would imply, among other things, that if G is planar and $\rho \ge 7$, then G is in C^1. So far we have that if G is planar and $\rho \ge 8$, then $G \in C^1$. This is proved by Vizing in [14].

It is not difficult to construct planar C^2_ρ-graphs for $2 \le \rho \le 5$; examples of these are C_{2k+1} and the following three Platonic graphs with a vertex introduced into any one of the edges: the tetrahedron, the octahedron and the icosahedron.

For $\rho = 7$ we have the following partial result:

Theorem 11. *If* $G \in C_7^2$ *and is planar, then there must be at least six vertices of maximal valency* 7.

Proof. Let ϕ_j denote the number of vertices of degree j and $\phi_7(i_1, i_2, i_3, i_4)$ denote the number of vertices of degree 7 adjacent to exactly i_1 vertices of degree 2, i_2 vertices of degree 3, i_3 vertices of degree 4, and i_4 vertices of degree 5. We have:

$$2\phi_2 = \phi_7(1, 0, 0, 0);$$
$$2\phi_3 \le \phi_7(0,1,0,0) + \phi_7(0,1,1,0) + \phi_7(0,1,0,1) + 2.\phi_7(0,2,0,0);$$
$$2\phi_4 \le \phi_7(0,1,1,0) + 2.\phi_7(0,0,2,0) + 3.\phi_7(0,0,3,0) + 2.\phi_7(0,0,2,1) + \phi_7(0,0,1,2) + \phi_7(0,0,1,1);$$
$$2\phi_5 \le \phi_7(0,0,0,1) + 2.\phi_7(0,0,0,2) + 3.\phi_7(0,0,o,3) + 4.\phi_7(0,0,0,4) + \phi_7(0,0,2,1) + 2.\phi_7(0,0,1,2) + \phi_7(0,1,0,1) + \phi_7(0,0,1,1),$$

$$\Rightarrow 4.\phi_2 + 3.\phi_3 + 2.\phi_4 + \phi_5 \le 3\sum_{(k)} \phi_7(i_1, i_2, i_3, i_4) \le 3.\phi_7 .$$

Also, Euler's theorem for polyhedra implies that:

$$\phi_7 + 12 \le \phi_5 + 2.\phi_4 + 3.\phi_3 + 4.\phi_2.$$

These last two statements together imply the required result. //

For $\rho = 6$, a result similar to the above can be stated as follows: If $G \in C_6^2$ and is planar, then G must have at least four vertices of maximum valency 6.

3. Connectivity in C*-graphs

We now deal with the question of the connectivity of C*-graphs. The constructions given also settle the problem of the existence of such graphs.

First we establish a couple of preliminary lemmas.

Lemma 2. *For* $i = 1, 2$, *let* G_i *be two* C_3^**-graphs consisting of an odd circuit* C_{2k+1} *and a set* S_i *of* k *independent edges. If* $S_1 \cap S_2 = \emptyset$, *then* $G = G_1 \cup G_2 \in C_4^*$.

Proof. $G \in C_4^2$ by Theorem 3 (i).

We have to show that the removal of any edge e yields a 4-colourable graph.

Case 1

$e \in C_{2k+1}$: Colour S_1 with colour α,
S_2 with colour β,
$C_{2k+1} \setminus \{e\}$ with colours γ, δ alternately.

Case 2

$e \in S_i$: By hypothesis, $G_i \setminus \{e\}$ is 3-colourable. Now, colour S_j $(j \neq 1)$ with the fourth colour. //

This result can be generalized as follows:

Lemma 3. Let $\{G_i\}_{i=1}^{\rho-3}$ be $(\rho - 3)$ C_3^*-graphs consisting of an odd circuit C_{2k+1} and a set $\{S_i\}_{i=1}^{\rho-3}$ of k independent edges. Then if $S_i \cap S_j = \emptyset$ for all i, j $(i \neq j)$, we have:

$$G_t = \bigcup_{i=1}^{t} G_i \in C_{t+2}^* \text{ for each } t \text{ satisfying } 2 \leq t \leq \rho - 3. \quad //$$

We should mention at this stage that a computer programme has been written jointly with C. Galea and A. Buttigieg which constructs C_ρ^*-graphs from C_t^*-graphs $(t < \rho)$ using the above-mentioned techniques.

Lemma 4. (Jakobsen, [8]) If $G_1, G_2 \in C_\rho^*$, then any Hajos union G of G_1 and G_2 also lies in C_ρ^*. //

We can now prove the following theorem:

Theorem 12. For each odd n and for each t satisfying $2 \leq t \leq (n - 4)$, there exists a C_t^*-graph of order n and having a vertex of valency 2.

Proof. The method of proof is by constructing $(n - 5)$ C_3^*-graphs $\{G_i\}_{i=1}^{n-5}$, each consisting of an odd circuit C_n and an independent set $\{S_i\}_{i=1}^{n-5}$ of $\frac{1}{2}(n - 1)$ edges satisfying $S_i \cap S_j = \emptyset$ for all i, j $(i \neq j)$.

Having constructed these graphs we can then apply Lemma 3. The C_3^*-graphs can be split into three families as follows: Let $n = 2k + 1$.

Then the first family consists of $(k-4)$ graphs $\{H_{1r}\}_{r=1}^{k-4}$ where the independent edges of H_{1r} are given by:

(1, r+2)

(r+2-i, r+2+i) (i = 1, 2, ..., r)

(2k, k+r), (2k-1, k+r+1)

(k+r-j, k+r+1+j) (j = 1,2,...,k-r-3)

(cfr. diagram)

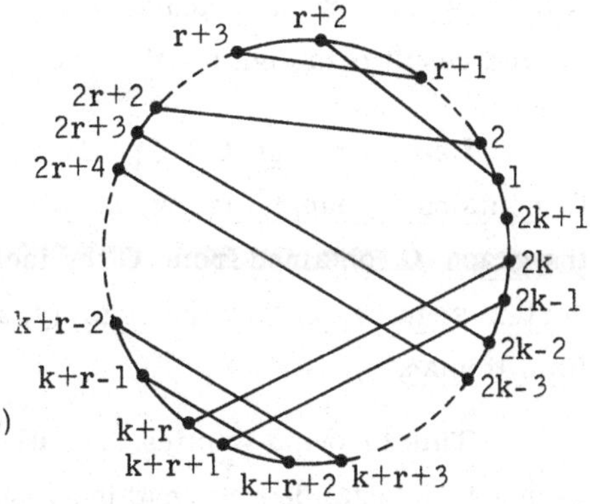

The second family also consists of $(k-4)$ graphs $\{H_{2r}\}_{r=1}^{k-4}$, where the independent edges of the rth graph are given by:

(1, k+r+1)

(k+r-i+1, k+r+i+1) (i=1, 2, ..., k-r-3)

(2k, r+2), (2k-1, r+3)

(r+2-j, r+3+j) (j=1, 2, ..., r)

(cfr. diagram)

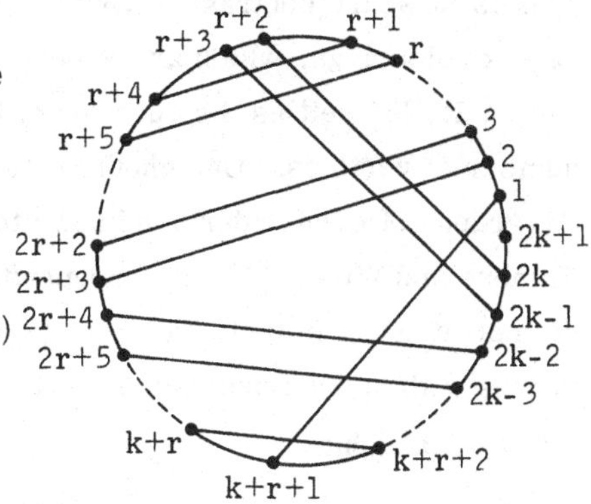

The last family consists of three graphs $\{H_{3r}\}_{r=1}^{3}$ whose respective independent edges are given by:

(i) (2k, k-1), (2k-1, k)

(k+i, k-1-i) (i=1, 2, ..., k-2)

(cfr. diagram)

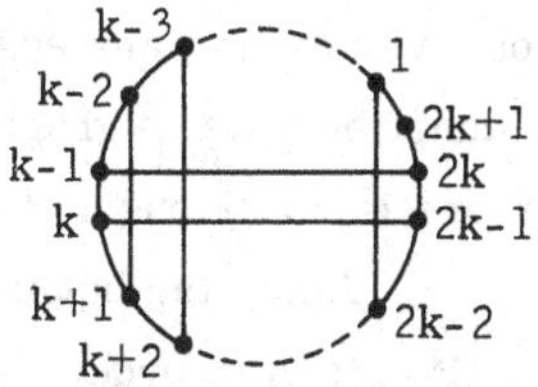

(ii) (k, 2k)

(k-i, k+i) (i=1, 2, ..., k-1)

(cfr. diagram)

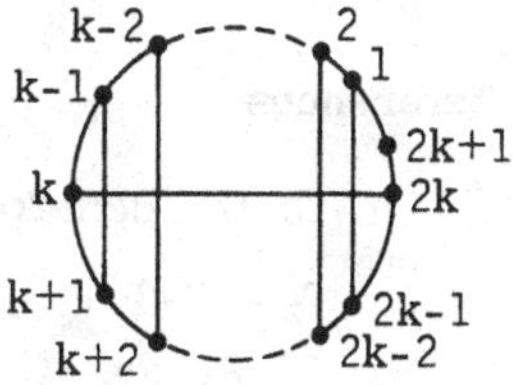

(iii) (1, k+1)

(k+1-i, k+1+i) (i=1, 2, ..., k-1)

(cfr. diagram)

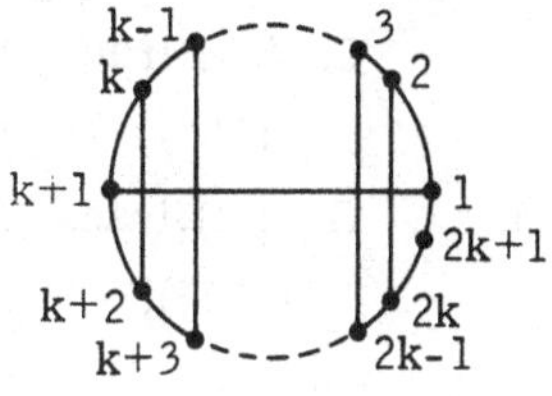

A tedious case-by-case analysis and repeated use of the following easily-proved result establish the theorem. //

Lemma 5. If $G \in C_3^*$, G has a Hamiltonian circuit C_{2k+1} and G contains a triangle (v, w, x) with (v, w) and $(w, x) \in C_{2k+1}$, then the graph $\tilde{G}$ obtained from G by inclusion of a vertex y in (v, w), a vertex z in (w, x) and the edge (y, z) satisfies: $\tilde{G} \in C_3^*$ and is Hamiltonian. //

This theorem implies the rather negative conclusion that one cannot in general deduce anything about the connectivity of C*-graphs. This is in sharp contrast to vertex-critical and edge-critical graphs in vertex colourings, which are at least ρ-edge-connected (cfr. [11]).

It also settles the question of the existence of C*-graphs on an odd number of vertices. One should note in this connexion, that to date no C*-graphs of even order are known to exist. It has been conjectured by Beineke and Wilson [1], and independently by Jakobsen [9], that these do not exist. In fact, Jakobsen [9], has shown that there are no C_3^*-graphs with an even number of vertices n, where n is at most 10.

4. Addendum

Since the conference, the first author has established that the graphs obtained from K_{2k} and K_{2k} with a 1-factor deleted by the insertion of a vertex into any edge lie in C_{n-2}^* and C_{n-3}^* respectively. As there cannot be a C_{n-1}^*-graph of order n, we see that for each odd integer n, there is a C_ρ^*-graph of order n, with minimal valency 2 for all possible ρ.

Also, the bound of Theorem 10 has been improved to the following: A C_ρ^*-graph of order n has at least $\frac{1}{4}n(\rho + 1)$ edges.

References

1. L. W. Beineke and R. J. Wilson. On the edge chromatic number of a graph. Discrete Math., 5, no. 1, (1973).
2. C. Berge. Graphes et hypergraphes. Dunod, Paris (1970).
3. G. A. Dirac. A note on the colouring of graphs. Math. Zeitschr., 54 (1951), 347-53.
4. G. A. Dirac. The structure of k-chromatic graphs. Fund. Math., 40 (1953), 42-55.

5. S. Fiorini and R. J. Wilson. On the chromatic index of a graph, I. Cahiers du Centre d'Etudes de Recherche Opérationnelle (15) 3 (1973), 253-62.

6. F. Harary. A seminar on graph theory. Holt, Rinehart and Winston, New York (1967).

7. F. Harary. Graph theory. Addison-Wesley, Reading, Mass., (1969).

8. I. T. Jakobsen. Some remarks on the chromatic index of a graph. Aarhus Preprint Series No. 40, 1971/72.

9. I. T. Jakobsen. On critical graphs with chromatic index 4. Aarhus Preprint Series No. 24, 1972/73.

10. G. H. J. Meredith. Regular n-valent, n-connected, non-hamiltonian, non-n-edge-colourable graphs. J. Combinatorial Theory, 14 (1973), 55-60.

11. O. Ore. The four color problem. Academic Press, New York (1967).

12. P. Turan. Eine Extremalaufgabe aus der Graphentheorie. Mat. Fiz. Lapok, 48 (1941), 436-52.

13. V. G. Vizing. On an estimate of the chromatic class of a p-graph. (Russian) Diskret. Analiz., 3 (1964), 25-30.

14. V. G. Vizing. Critical graphs with a given chromatic class. (Russian) Diskret. Analiz., 5 (1965), 9-17.

15. V. G. Vizing. The chromatic class of a multigraph. Cybernetics, 1, no. 3, (1965), 32-41.

16. V. G. Vizing. Some unsolved problems in graph theory. Russian Math. Surveys, 23 (1968), 125-42.

17. R. J. Wilson. Introduction to graph theory. Oliver and Boyd, Edinburgh (1972).

The Open University,
Milton Keynes, England

ON A THEOREM OF R. A. LIEBLER

A. GARDINER

The conjecture that 'any affine plane **A** admitting a collineation group which is rank 3 on points is a translation plane' was stated by D. G. Higman [3]; M. J. Kallaher [4] proved a slightly weaker statement namely that **A** is either a translation plane or the dual of a translation plane, and R. A. Liebler [5] overcame the possible ambiguity to prove the conjecture. We offer a proof which, though similar to [4] and [5] in the first stages, completes the final steps more smoothly. Thus we prove

Theorem (Liebler). An affine plane **A** admitting a collineation group H which is rank 3 on points is a translation plane, (and H contains all translations).

Let **A** be an affine plane of order k, with projective completion **P** having line at infinity $l(\infty)$. $P(\infty)$, $P(1)$, ..., $P(k)$ are the points of $l(\infty)$ and $l(1)$, ..., $l(k)$ the affine lines through $P(\infty)$. For a point P and line l of **P**, (P) denotes the set of lines through P and (l) the set of points on l. If G is a group acting on the set Ω, then G^{Ω} is the natural permutation group induced by G on Ω.

Note that if G^{Ω} is transitive, $|\Omega| = m$, and B is a subgroup of G of index relatively prime to m, then B^{Ω} is transitive. We use the following results.

I [8]. An affine plane admitting a group of collineations which is transitive on affine flags is a translation plane.

II [2]. A collineation of order two of a projective plane is an elation, a homology or a Baer involution.

III [5]. Suppose G^{Ω} is transitive, $P \in \mathrm{Syl}_p(G)$ for some prime p. If $h \in Z(P)$ and $\Lambda = \{\alpha \in \Omega : \alpha h = \alpha\}$, then either $\Lambda = \phi$, or $|\Omega|_p \,\big|\, |\Lambda|$.

IV (For example, [6] proof of Theorem 1). Let l be a line of the affine plane **A**, and X a collineation group fixing l with $X^{(l)}$ doubly transitive. Then an involution in the centre of a Sylow 2-subgroup of $X^{(l)}$ is not a Baer involution.

V [7]. Let **A** be an affine plane of order k, with collineation group H. If k of the points at infinity are centres of non-trivial translations in H, then so is the other point.

VI [1]. Let **A** be an affine plane with collineation group H transitive on the points of $(l) - l \cap l(\infty)$. If H contains a homology with axis $l(\infty)$ and centre on l, then H contains all translations in the direction $l \cap l(\infty)$.

All but I and II have very simple proofs. From now on **A** is an affine plane with collineation group H acting as a rank 3 group on affine points. In view of I we assume that H is not transitive on affine flags.

Proposition 1. For each affine point P, H_P has two orbits Δ_P, Γ_P on (P) determined by the two orbits Δ, Γ of H on $(l(\infty))$.

Proof. Since H is a rank 3 group on affine points, H_P has at most two orbits on (P). If H_P is transitive on (P), then H is transitive on affine flags. If H is transitive on $(l(\infty))$, then $|H : H_P| = k^2$ is relatively prime to $(k + 1)$, so H_P is transitive on $(l(\infty))$. //

Proposition 2. For each affine line $l \in (P(i))$, $(H_l)^{(l)-P(i)}$ is doubly transitive.

Proof. For $P \in (l) - P(i)$, H_P is transitive on the affine points of lines in each of Δ_P and Γ_P. Thus $H_{P,l}$ is transitive on $(l) - \{P, P(i)\}$. Since we may assume $k \geqq 3$, the assertion follows. //

Proposition 3. Either H fixes a point of $l(\infty)$, $P(\infty)$ say, or **A** is a translation plane.

Proof. For each affine line $l \in (P(i))$, $(H_l)^{(l)-P(i)} = G$ is doubly transitive, so the translations of **A** in the direction P(i) form a normal subgroup of G which is either trivial or transitive (and hence regular).

Suppose $|\Delta| \neq 1 \neq |\Gamma|$. It suffices to produce a single non-trivial translation. Let $S \in Syl_2(G)$ and $h \in Z(S)$ be of order two. Then if k is even, h is a translation in the direction $P(i)$. If k is odd, h is a homology fixing l, $l(\infty)$, $P(\infty)$, $P(i)$ and some affine point $P \in (l)$; let $l(j) = PP(\infty)$. If h has axis $l(\infty)$, we are done by VI. Otherwise h has axis $l(j)$ and centre $P(i)$, so we are done by the dual of VI. //

From now on we assume that H fixes $P(\infty)$. Let C be the kernel of the action of H on $(l(\infty))$ and D the kernel of the action of H on $(P(\infty))$; set $K = C \cap D$, the group of translations in H of $\mathbf{A}$ in the direction $P(\infty)$.

Proposition 4. (i) D is non-trivial if and only if K is non-trivial if and only if C is non-trivial.

(ii) If $K \neq 1$, then K is an elementary abelian p-group of order $k = p^r$, for some prime p, acting regularly on $(l(i)) - P(\infty)$, $i = 1, \ldots, k$.

Proof. (i) Assume $D \neq 1$; (the corresponding argument for $C \neq 1$ is completely dual). Each $d \in D - 1$ is a central collineation with axis $l(\infty)$ and centre P. If $P \in (l(\infty))$, then either (a) $P = P(\infty)$, so $K \neq 1$, or (b) $P = P(i)$ and the transitivity of H on $(l(\infty)) - P(\infty)$ yields non-trivial translations in each direction $P(i)$, $1 \leq i \leq k$, and so also in the direction $P(\infty)$, by V, whence $K \neq 1$. If $P \notin (l(\infty))$, then $\mathbf{A}$ is a translation plane by VI.

(ii) K is a normal subgroup of $H_{l(i)}$, so $K \neq 1$ implies that K is regular on $(l(i)) - P(\infty)$, by Proposition 2. For $P \in (l(i)) - P(\infty)$, H_P acts transitively on $K - 1$ by conjugation, so K is elementary abelian. //

Proposition 5. For $1 \leq i, j \leq k$, $H_{P(i), l(j)}$ acts doubly transitively on both $(P(i)) - l(\infty)$ and $(l(j)) - P(\infty)$.

Proof. Since H_P is transitive on $(l(\infty)) - P(\infty)$, $H_{P(i)}$ is transitive on affine points; thus $H_{P(i), l(j)}$ acts transitively on $(P(i)) - l(\infty)$ and on $(l(j)) - P(\infty)$; since $|H : H_{P(i), l(j)}| = k^2$ is relatively prime to $(k - 1)$, $H_{P(i), l(j)}$ is doubly transitive as asserted. //

Proposition 6. A <u>is a translation plane.</u>

Proof. Let $i \neq \infty$ and $l \in (P(i)) - l(\infty)$. Set $(H_l)^{(l)} = G$ and choose $S \in Syl_2(G)$, $h \in Z(S)$ of order two. If k is even, then h is a translation with centre P(i), whence **A** is $(l(\infty), P(i))$-transitive, $1 \leq i \leq k$, and so is a translation plane. If k is odd, h is a homology with axis either (a) $l(\infty)$, or (b) $l(j)$, $j \neq \infty$. In case (a) h has affine centre so we are done by VI. In case (b) **A** is a translation plane by the dual of VI as in the proof of Proposition 3. //

References

1. J. André. Über Perspektivitäten in endlichen projektiven Ebenen. Arch. der Math., 6 (1954), 29-32.
2. R. Baer. Projectivities with fixed points on every line of the plane. Bull. Am. Math. Soc., 52 (1946), 273-86.
3. D. G. Higman. On finite affine planes of rank 3. Math. Z., 104 (1968), 147-9.
4. M. J. Kallaher. On finite affine planes of rank 3. J. Algebra, 13 (1969), 544-53.
5. R. A. Liebler. Finite affine planes of rank three are translation planes. Math. Z., 116 (1970), 89-93.
6. T. G. Ostrom and A. Wagner. On projective and affine planes with transitive collineation groups. Math. Z., 71 (1959), 186-99.
7. F. Piper. Elations of finite projective planes. Math. Z., 82 (1963), 247-58.
8. A. Wagner. On finite affine line transitive planes. Math. Z., 87 (1965), 1-11.

University of Birmingham
Birmingham, England

OUTERTHICKNESS AND OUTERCOARSENESS OF GRAPHS

RICHARD K. GUY

0. Introduction

The thickness [1, 3, 4, 10, 11] and coarseness [2, 8] of some of the more frequently studied families of graphs have been calculated or estimated; the analogous concepts of the title are mentioned in [5] but their values do not appear to have been calculated for these families. So the complete graph, K_n, the complete bipartite graph, $K_{m,n}$, and the (1-skeleton of the) n-dimensional cube, Q_n, provide six problems which are considered here. Proofs and further results will appear elsewhere in an extended version of this paper.

1. Definitions

A planar graph is one which can be imbedded in the plane or sphere. The thickness (coarseness) of a graph is the least (greatest) number of parts in an edge-partition of the graph into planar (non-planar) subgraphs.

When a graph is imbedded in a surface, the connected components of the complement with respect to that surface are called faces. An outerplanar graph is a (planar) graph admitting a plane imbedding with all the vertices on the boundary of a single face.

The outerthickness, θ_0, and outercoarseness, ξ_0, of a graph are the analogs of thickness and coarseness with 'planar' replaced by 'outer-planar'.

2. The outerthickness of the complete graph

A maximal outerplanar graph is a triangulation of a polygon; such a graph on n points contains $2n - 3$ edges. The complete graph on n points contains $\binom{n}{2}$ edges, so

$$\theta_0(K_n) \geq \{\frac{n(n-1)}{2(2n-3)}\} = [\frac{n}{4}] + 1 \quad (n > 2)$$

where braces (brackets) denote 'least (greatest) integer not less (greater) than'. Geller [7] has shown the inequality to be strict when $n = 7$, but we believe equality holds for all other $n \geq 3$. We have proved this for $n \leq 23$, $n = 2^k$ and $n = 2^k + 1$.

3. The outercoarseness of the complete graph

Halin [9, see also 6, 7] has characterized outerplanar graphs as those having no subgraph homeomorphic to K_4 or $K_{3,2}$. These each have 6 edges, so

$$\xi_0(K_n) \leq [\frac{n(n-1)}{12}].$$

We can show equality for all $n \geq 1$.

4. The outerthickness of the complete bipartite graph

A maximal outerplanar bipartite graph on $m + n$ points $(m \leq n)$ contains $2m + n - 2$ edges, and the complete bipartite graph contains mn edges; also $K_{m,n}$ can be partitioned into m copies of $K_{1,n}$, which is outerplanar, so

$$m \geq \theta_0(K_{m,n}) \geq \{\frac{mn}{2m + n - 2}\} \quad (m \leq n).$$

In general, $\theta_0(K_{m,n})$ is not equal to either of these bounds. The following results are known:

$$\theta_0(K_{m,n}) = \begin{cases} m & (n > m(m-1)), \\ m-1 & (m(m-1) \geq n > (m-1)(m-2)), \\ m-2 & ((m-1)(m-2) \geq n > \frac{2}{3}(m-1)(m-3)), \\ 2 & (m = n = 5), \\ 3 & (m = 6 \leq n \leq 10;\ m = 7 \leq n \leq 9;\ m = n = 8), \\ 4 & (m = 7,\ 10 \leq n \leq 16;\ m = 8,\ 9 \leq n \leq 14). \end{cases}$$

5. The outercoarseness of the complete bipartite graph

$$\xi_0(K_{m,n}) \leq [\frac{mn}{6}].$$

Equality holds, except that $\xi_0(K_{1,n}) = 0$ for all n.

6. **The outerthickness of the n-cube**

$$\theta_0(Q_n) \geq \left\{ \frac{n.\, 2^{n-1}}{3.\, 2^{n-1} - 2} \right\} = [\tfrac{n}{3}] + 1 \quad (n \geq 1).$$

Equality holds for all $n \geq 1$.

7. **The outercoarseness of the n-cube**

A subgraph of Q_n, homeomorphic to $K_{2,3}$, contains at least 8 edges, and one homeomorphic to K_4 at least 9, so

$$\xi_0(Q_n) \leq [n.\, 2^{n-1}/8] = [n.\, 2^{n-4}] \quad (n \geq 0).$$

Equality holds for $n \leq 3$, but not in general; $\xi_0(Q_4) = 3$.

References

1. J. Battle, F. Harary and Y. Kodama. Every planar graph with nine points has a non-planar complement. Bull. Amer. Math. Soc., 68 (1962), 569-71; MR 27 #5248.
2. L. W. Beineke and R. K. Guy. The coarseness of $K_{m,n}$. Canad. J. Math., 21 (1969), 1086-96; MR 41 #6727.
3. L. W. Beineke and F. Harary. The thickness of the complete graph. Canad. J. Math., 17 (1965), 850-9; MR 29 #1636.
4. L. W. Beineke, F. Harary and J. W. Moon. On the thickness of the complete bipartite graph. Proc. Cambridge Philos. Soc., 60 (1964), 1-5; MR 28 #1611.
5. G. Chartrand, D. Geller and S. Hedetniemi. Graphs with forbidden subgraphs. J. Combinatorial Theory, 10B (1971), 12-41; MR 44 #2645.
6. G. Chartrand and F. Harary. Planar permutation graphs. Ann. Inst. H. Poincaré, Sec. B3 (1967), 433-8; MR 37 #2626.
7. D. P. Geller. Outerplanar graphs [see F. Harary. Graph Theory, Adison-Wesley, 1969, pp. 108, 245].
8. R. K. Guy and L. W. Beineke. The coarseness of the complete graph. Canad. J. Math., 20 (1966), 888-94; MR 37 #2633.

9. R. Halin. Über einen graphentheoretischen Basisbegriff und seine Anwendung auf Färbensprobleme. Doctoral thesis, Köln, 1962.

10. M. Kleinert. Die Dicke des n-dimensionalen Würfel-Graphen. J. Combinatorial Theory, 3 (1967), 10-15; MR 35 #2776.

11. W. T. Tutte. On the non-biplanar character of the complete 9-graph. Canad. Math. Bull., 6 (1963), 319-30; MR 28 #2535.

The University of Calgary,
Alberta, Canada

GRAPHS WITH HOMEOMORPHICALLY IRREDUCIBLE SPANNING TREES

ANTHONY HILL*

By a graph G we mean here a linear network, at least 3-connected with no vertices of degree 2, no multiple edges and no loops. A graph T is called a spanning graph of G provided (i) T is a tree, (ii) T is a subgraph of G and (iii) every vertex of G belongs to T.

A graph is said to be homeomorphically irreducible (HI) if it has no nodes of degree 2. Thus an HI tree may be a spanning tree of G and we shall call this graph a HISTree of G.

Starting with the example of the set of all 3-polytopes, inspection shows that while all possess spanning trees [1] not all have HISTrees. (1) We would like to know which graphs possess HISTrees, if such graphs are common and what may be found as contingent with the existence or non-existence of this feature.

1. Types of HISTrees

For the purpose of the present note it is sufficient to regard these graphs as belonging to three main sub-species. These are:

(i) Stars (as in fig. 4, which shows the smallest).

(ii) Star chains (as in figs. 5, 6.1 and 10.1).

(iii) Star trees (as in fig. 7.1).

2. Types of 3-polytopes with HISTrees

The most obvious example of a family of graphs G containing a HISTree is the family of those 3-polytopes in which those edges not belonging to the HISTree comprise a circuit joining the terminal vertices of the HISTree. This family of 3-polytopes may also be defined by the following and equivalent definitions:

---

* Research supported by a Leverhulme Fellowship.

(i) G has F faces and one face (that formed by the circuit described above) has F - 1 edges.

(ii) The face containing F - 1 edges shares an edge with all of the remaining faces of G.

Examples of this family of 3-polytopes are shown in figs. 4, 5, 12.1, 12.2.

The simple (i. e. trivalent) variety of the family of G described seems first to have been discussed by Kirkman [2] who attempted an enumeration. Much later Rademacher [3] surveyed the topic and offered solutions to the problem of enumeration. If we remove the restriction of trivalent vertices for the HISTree we have a complete family of 3-polytopes which remain covered by the definitions already given. The question of enumeration is also basically the same, it asks for the number of unlabelled combinatorially distinct planar HI trees; a problem probably already solved. In the terminology of Kirkman and Rademacher this larger family comprises 'degenerate' versions of the trivalent kind which Rademacher calls 'based'. A term for these degenerate kind (in fact the whole family) was suggested to Rademacher by Polya is 'roofless'. However rather than use this term, or offer that of Polyatopes - which might be confusing - we shall call them r-topes.

Concerning r-topes we make the following observations:

(i) All r-topes are Hamiltonian[(2)] (containing an adjacent Hamilton circuit) [4].

(ii) Simple r-topes can have at most 3 topologically distinct Hamilton circuits [4].

(iii) If the embedding of the HISTree admits no symmetry in the plane, then the resulting r-tope will have no symmetry (i. e. it will be of group order 1) [5].

(iv) With the exception of the tetrahedron and the triangular prism, (figs. 4 and 5) we can find no other examples of simple 3-polytopes that are vertex transitive which contain a HISTree.

(v) If a simple 3-polytope contains a HISTree then the remaining edges may comprise a circuit such that it is an r-tope; if, however, they do not comprise a single circuit then they will be found to comprise n circuits joining at least 3 terminal vertices in each. We show the smallest versions of these in figs. 6 and 7.

Concerning the general case of G with a HISTree, other than r-topes, we look first at those which are planar.

(i) Simplicial 3-polytopes[3]: We conjecture that they always contain a HISTree. However not all simplicial 3-polytopes contain an r-tope as a spanning graph. Fig. 9.1 is the smallest, having 8 vertices, which does not admit a circuit, but it does possess a path which connects the terminal vertices of the HISTree.

(ii) 3-polytopes (other than those described) may often have an r-tope as a spanning graph. Equally they may only possess a path, e.g. fig. 8. Finally, a 3-polytope may possess neither circuit or path, e.g. fig. 9.2.

(iii) Of the 22 3-polytopes with $E \leq 12$, only three are neither r-topes nor contain an r-tope as a spanning graph. These are shown in figs. 1, 2 and 3.

Of the non-planar G with HISTrees we observe the following:

(i) the complete graph on n vertices, K_n, is alone in containing all the HISTrees on n vertices, and these occur as r-topal spanning graphs.

(ii) numerous examples of n-dimensional polytopes are seen to have HISTrees.

(iii) Möbius Ladders [6] on n even vertices are seen to be a version of the r-tope except that the HISTree undergoes an immersion. We shall call such graphs r-cradles. In figs. 15 to 18 we show some examples of Möbius Ladders and the invariant manner in which the HISTree grows, as n grows larger.

(iv) The Thompson graph (fig. 15), the Petersen graph and the Heawood graph, all n-cages, are seen to be r-cradles.

(v) The problem of enumerating all topologically distinct r-cradles would seem to be hard.

It is to be noted that a HISTree can give rise to a number of topologically distinct r-cradles by their successive immersion to isomorphism. It would be instructive to enumerate the topologically distinct r-cradles derived from immersion of a single abstract HI tree.

Excluding the four r-topes obtained from all distinct embeddings of the two cubic HI trees on 12 vertices (these may be called the Frucht trees), of the remaining 43 3-connected cubic graphs on 12 vertices 39 are r-cradles (all derived from the above mentioned two trees).

(vi) It is open to conjecture that all r-cradles contain a Hamiltonian path, as does the Petersen graph which has no Hamilton circuit.

3. Some constructions with HISTrees

We have mentioned the method of obtaining graphs of group order 1 from suitable HItrees, this method may be seen to be the result obtained when we join each successive terminal edge of a HI tree to a 'mid-point' on the adjacent edge, moving in a clockwise or anti-clockwise direction. We see that fig. 12.1 is isomorphic with fig. 12.2. In this manner any HI tree can be 'tied up' to form an r-tope.

Taking the same HI tree (the smallest of its kind with respect to symmetry), we can perform a different kind of 'tying up' and obtain asymmetric 3-polytopes. Figs. 10.1 and 11.1 show how the same tree can undergo two distinct 'combings' such that when the terminal edges are attached to a quadrangular circuit (a 4-gon) the result in each case is an asymmetric 3-polytope. Figs. 10.3 and 11.3 are in fact the two smallest asymmetric 3-polytopes; they are each self-dual and each possesses an r-topal spanning graph.

The same construction, choosing a cubic tree which has been given an asymmetric embedding in the plane is shown in fig. 13.1 and 13.2, results in a 3-polytopal graph with no symmetry.

A different construction applied to a 'star tree' HI tree, figs. 14.1-14.3, results in an asymmetric graph, 3-polytopal, on 8 vertices.

The two cubic HI trees, fig. 13.1 and fig. 14.1, may be called the Frucht trees since they give rise to the two Frucht G. They are the smallest simple 3-polytopes (they are of course r-topes) with no symmetry, [7] [8]. The graph shown in Fig. 7.4, also based on a Frucht tree, is asymmetric.

References

1. D. W. Barnette. Trees in polyhedral graphs. Canadian J. Math., 18 (1966).
2. T. P. Kirkman. On the enumeration of x-edra having trihedral summits and on (x - 1) ground base. Phil. Trans. R. Soc. (1856).
3. H. Rademacher. On the number of certain types of polyhedra.

Illinois J. Math., vol. 9 (1966).
4. A. Hill and C. A. Rogers. Some Hamiltonian graphs. J. Combinatorial Theory (to appear).
5. A. Hill. Some problems from the visual arts. New York Acad. Sciences, Vol. 175, Internat. Conf. on Combinatorial Math. N. Y., July 1970.
6. R. K. Guy and F. Harary. On the Möbius Ladders. Canadian Math. Bull., 10 (1967).
7. R. Frucht. Graphs of degree three with a given abstract group. Canadian J. Math., 1 (1949).
8. A. T. Balaban, R. O. Davies, F. Harary, A. Hill and R. Westwick. Cubic identity graphs and planar graphs derived from trees. J. Australian Math. Soc., Vol. XI, part 2 (1970).

University College,
London

Addendum

1. Can a polyhedral graph admit more than one line-disjoint HISTree? The answer is yes. We can show that the icosahedron admits two such HISTrees. We conjecture that the latter is the smallest graph where this is possible.
2. The authors expect to prove the following stronger theorem: for a polyhedral graph to be non-Hamiltonian it must have at least one pair of disjoint faces.
3. Does there exist a pair of dual polyhedral graphs such that neither possesses a HISTree? We have not so far found an example. But if a polyhedral graph is self-dual we can show that it may contain this feature. Let $C_n \times P_m$ be the cartesian product of a cycle and a path, and $C_n \times P^*_m$ be the result of contracting the inner cycle to a point. Then $C_4 \times P^*_3$ has no HISTree, nor does $C_5 \times P^*_3$. These two - possibly unique - cases have been shown to exist by Allen Schwenk (private communication to the author).

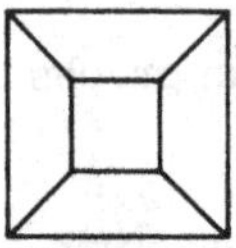

Fig. 1

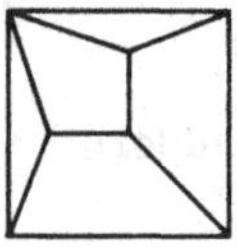

Fig. 2

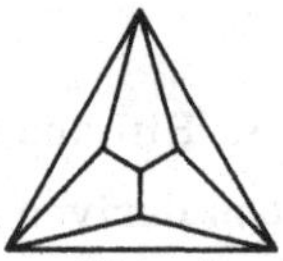

Fig. 3

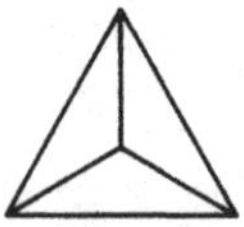

Fig. 4

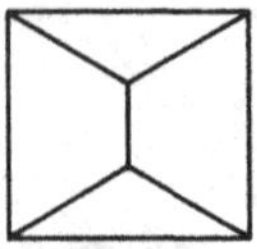

Fig. 5

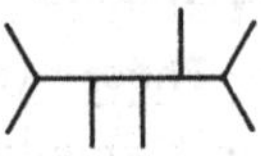

Fig. 6.1

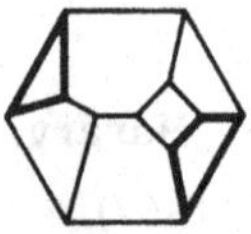

Fig. 6.2

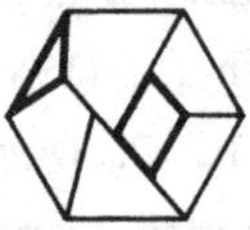

Fig. 6.3

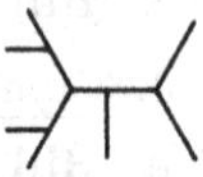

Fig. 7.1

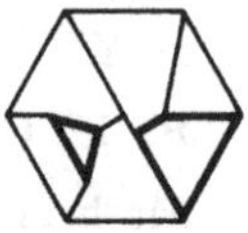

Fig. 7.2

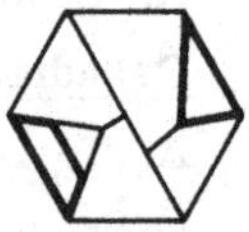

Fig. 7.3

Fig. 7.4

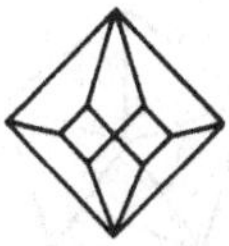
Fig. 8

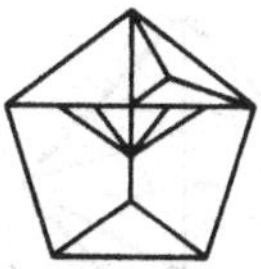
Fig. 9. 2

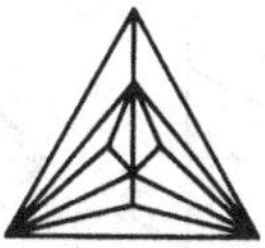
Fig. 9. 1

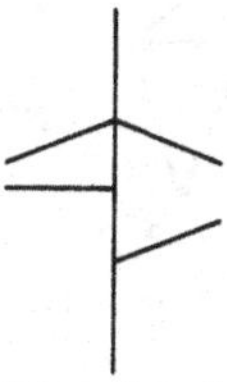
Fig. 10. 1

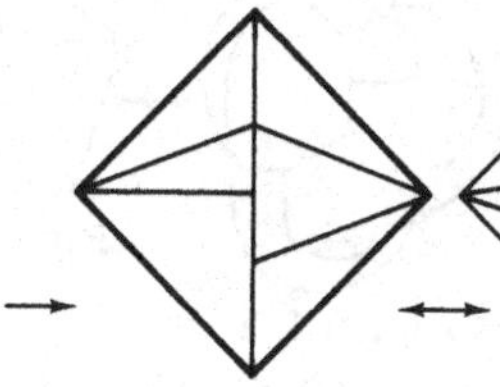
Fig. 10. 2

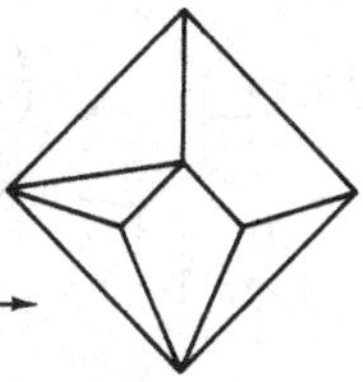
Fig. 10. 3

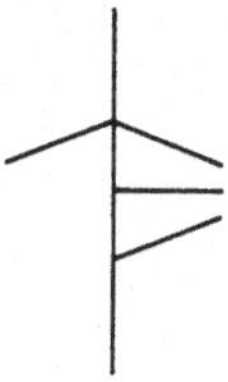
Fig. 11. 1

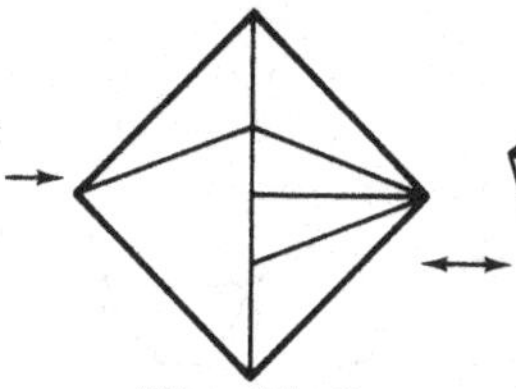
Fig. 11. 2

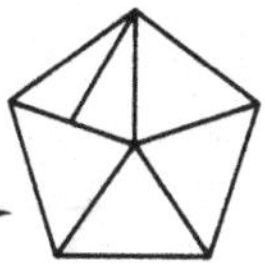
Fig. 11. 3

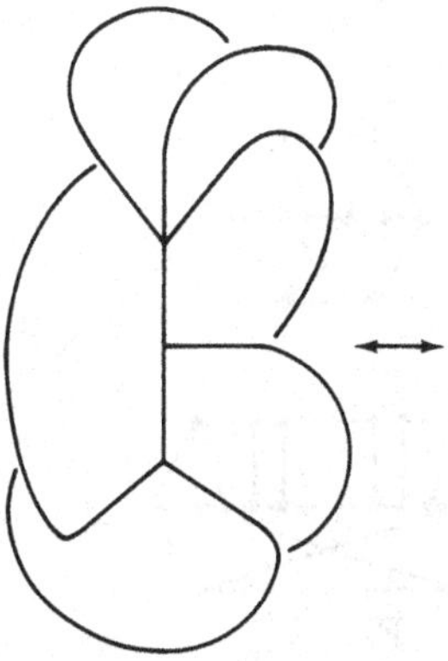
Fig. 12. 1

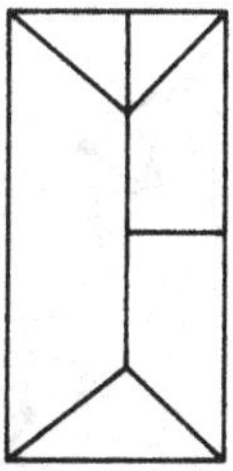
Fig. 12. 2

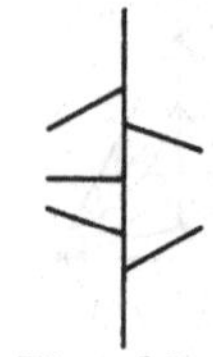
Fig. 13. 1

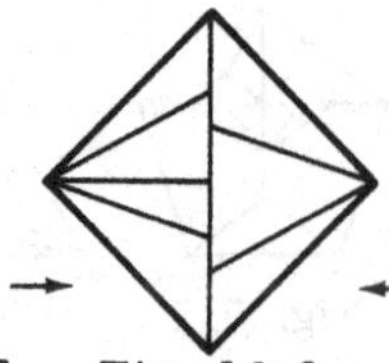
Fig. 13. 2

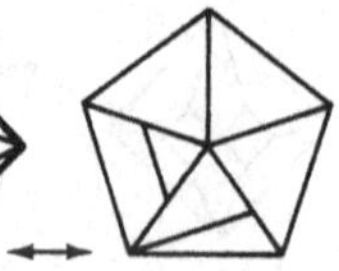
Fig. 13. 3

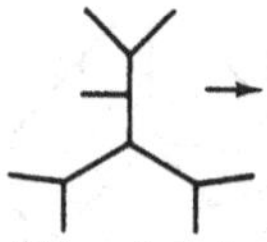
Fig. 14. 1

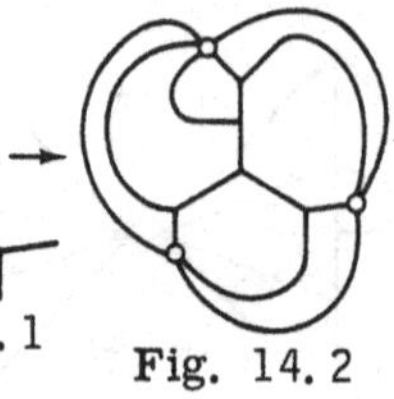
Fig. 14. 2

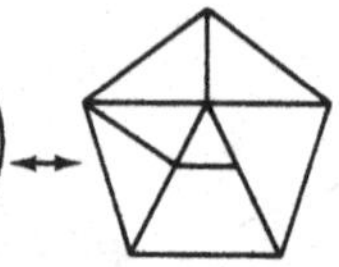
Fig. 14. 3

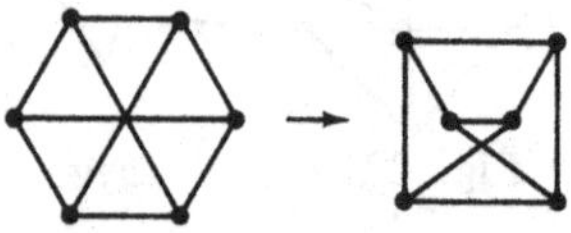
Fig. 15

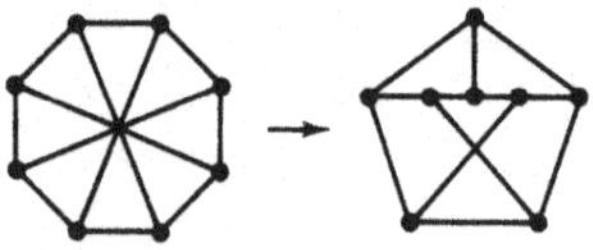
Fig. 16

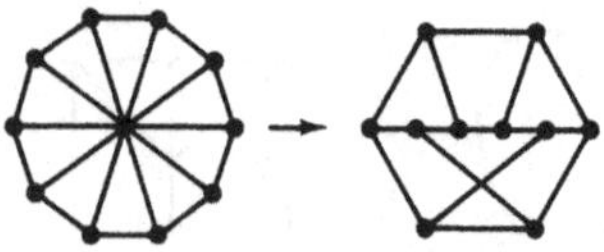
Fig. 17

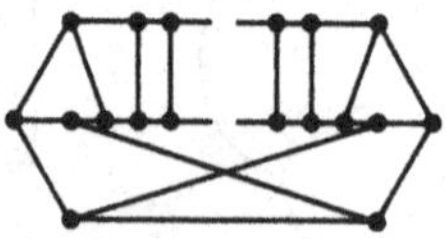
Fig. 18

A NOTE ON EMBEDDING LATIN RECTANGLES

A. J. W. HILTON

1. Introduction

The well-known theorem of H. J. Ryser [12] giving necessary and sufficient conditions for an $r \times s$ latin rectangle on $1, \ldots, n$ to be embedded in an $n \times n$ latin square on $1, \ldots, n$ was used by T. Evans [2] (and independently by S. K. Stein) to show that an incomplete $n \times n$ latin square on $1, \ldots, n$ can be completed to a $2n \times 2n$ latin square on $1, \ldots, 2n$. A similar, but somewhat more complicated, pair of theorems concerning symmetric latin squares was proved by A. Cruse [1].

The purpose of this note is to give alternative and, in my opinion, simpler proofs of the theorem of Ryser and the analogous theorem of Cruse. Ryser's theorem generalizes M. Hall's theorem [3] that an $r \times n$ latin rectangle on $1, \ldots, n$ can be embedded in an $n \times n$ latin square on $1, \ldots, n$, but the methods of proof seem to be rather dissimilar. The proof of Ryser's theorem which is given here is very obviously a simple generalization of the original proof of M. Hall's theorem.

There are still some open problems in this area (see [5], [7]), so it is possible that the existence of these alternative proofs may help towards the solution of some of these problems.

2. Definitions and a preliminary result

An $r \times s$ <u>latin rectangle</u> on $1, \ldots, n$ is a matrix of r rows and s columns, in which each entry is an integer between 1 and n, and which has the property that no integer occurs twice in any row or column. An $n \times n$ latin rectangle on $1, \ldots, n$ is called a <u>latin square</u> of order n on $1, \ldots, n$. An $r \times r$ latin rectangle $A = (a_{ij} : i, j = 1, \ldots, r)$ is <u>symmetric</u> if $a_{ij} = a_{ji}$ for all $i, j \in \{1, \ldots, r\}$.

A subset T of a set E is a transversal of a family $\mathfrak{A} = (A_\nu : \nu \in N)$ of subsets of E if there is a bijection $\psi : T \to N$ such that $x \in A_{\psi(x)}$ for all $x \in T$.

The following theorem is due to Hoffman and Kuhn [6]; L. Mirsky has shown ([9], [10]) by very neat arguments that it may be obtained from P. Hall's theorem [4].

Theorem 1. Let $\mathfrak{A} = (A_\nu : \nu \in N)$ be a finite family of subsets of a given set E, and let $M \subseteq E$ be given. Then $\mathfrak{A}$ has a transversal containing M if and only if the conditions (1) and (2) below are satisfied.

(1) $\left| \bigcup_{\nu \in N'} A_\nu \right| \geq |N'|$ for each $N' \subseteq N$.

(2) $|\{\nu \in N : M' \cap A_\nu \neq \phi\}| \geq |M'|$ for each $M' \subseteq M$.

3. Ryser's theorem

Let H be an $r \times s$ latin rectangle on $1, \ldots, n$. For each $i \in \{1, \ldots, n\}$ let $P(i)$ be the number of times i occurs in H. Then Ryser's theorem is as follows.

Theorem 2. The latin rectangle H can be embedded in a latin square K of order n on $1, \ldots, n$ if and only if

$P(i) \geq r + s - n$ for each $i \in \{1, \ldots, n\}$.

Proof. 1. Necessity. Suppose H is embedded in K. We may suppose $K = \begin{pmatrix} H & \vdots & X \\ \cdots & \cdots & \cdots \\ & Y & \end{pmatrix}$ where H, X, Y are $r \times s$, $r \times (n - s)$ and $(n - r) \times n$ matrices respectively. An integer $i \in \{1, \ldots, n\}$ occurs n times in K, $n - r$ times in Y, at most $n - s$ times in X, and therefore at least $n - (n - r) - (n - s) = r + s - n$ times in R.

2. Sufficiency. We first show by induction on s that H can be extended to an $r \times n$ latin rectangle on $1, \ldots, n$.

Let $N = \{1, \ldots, r\}$, and, for $\nu \in N$, let A_ν be the set of integers from $1, \ldots, n$ which do not occur in the νth row of H. Let $M = \{m_1, \ldots, m_l\}$ be the set of $i \in \{1, \ldots, n\}$ which occur exactly

$r + s - n$ times in H; we call the elements of M marginal elements. If we can show that the family $\mathfrak{A} = (A_\nu : \nu \in N)$ has a transversal which includes M, then it follows that there is an $r \times (s+1)$ latin rectangle on $1, \ldots, n$ in which each integer i, $1 \leq i \leq n$, occurs at least $r + (s+1) - n$ times, and so it then follows by induction that R can be extended to an $r \times n$ latin rectangle on $1, \ldots, n$.

First we show that $\mathfrak{A}$ satisfies condition (1) of Theorem 1. Let $N' \subseteq N$ and, for each $\nu \in N$, let $A_\nu = \{a_{\nu,1}, \ldots, a_{\nu,n-s}\}$. Consider the matrix

$$\mathbf{A} = (a_{\nu\mu} : \nu \in N', \mu \in \{1, \ldots, n-s\}).$$

The number of distinct elements occurring in $\mathbf{A}$ (i.e. $|\bigcup_{\nu \in N'} A_\nu|$) is at least the total number of entries in $\mathbf{A}$ (i.e. $(n-s)|N'|$) divided by the maximum number of times any element can occur as an entry (i.e. $r - (r+s-n) = n - s$). In other words,

$$\Big|\bigcup_{\nu \in N'} A_\nu\Big| \geq \frac{(n-s)|N'|}{(n-s)} = N'.$$

Therefore $\mathfrak{A}$ does satisfy condition (1) of Theorem 1.

Next we show that $\mathfrak{A}$ satisfies condition (2) of Theorem 1. Each marginal element is in exactly $r - (r+s-n) = n - s$ of the A_ν's, so for each $\alpha \in \{1, \ldots, l\}$ we may let

$$B_\alpha = \{A_\nu : \nu \in N, m_\alpha \in A_\nu\} = \{b_{\alpha,1}, \ldots, b_{\alpha,n-s}\}.$$

Consider the matrix

$$\mathbf{B} = (b_{\alpha\beta} : \alpha \in \{1, \ldots, l\}, m_\alpha \in M'; \beta \in \{1, \ldots, n-s\}).$$

The number of distinct elements occurring in $\mathbf{B}$ (i.e. $|\{\nu \in N : M' \cap A_\nu \neq \phi\}|$) is at least the total number of entries (i.e. $|M'| \times (n-s)$) divided by the maximum number of times any element can occur as an entry (which is the maximum number of marginal elements which may possibly be contained by any A_ν, i.e. $n - s$). In other words

$$|\{\nu \in N : M' \cap A_\nu \neq \phi\}| \geq \frac{|M'|(n-s)}{(n-s)} = |M'|.$$

Therefore $\mathcal{A}$ also satisfies condition (2) of Theorem 1.

By induction it now follows that H can be extended to an $r \times n$ latin rectangle on $1, \ldots, n$.

The extension of H which we have just described works equally well when $r = n$, so that we have proved that an $n \times s$ latin rectangle on $1, \ldots, n$ may be extended to an $n \times n$ latin square on $1, \ldots, n$ (the condition $P(i) \geq r + s - n$ is obviously satisfied in this case). Similarly an $n \times r$ latin rectangle on $1, \ldots, n$ may be extended to an $n \times n$ latin square on $1, \ldots, n$. This observation concludes the proof of Ryser's theorem.

Comment. When $r = n$, in the proof above, each element is a marginal element. In general, if $M = \bigcup_{\nu \in N} A_\nu$, to require that $\mathcal{A}$ has a transversal which includes M is the same as to require that $\mathcal{A}$ has a transversal. As is well-known $\mathcal{A}$ has a transversal if and only if condition (1) of theorem 1 is true. This is in fact a consequence of theorem 1 (put $M = \phi$). It is also a consequence of theorem 1 that when $M = \bigcup_{\nu \in N} A_\nu$ and condition (1) is true then condition (2) is true also. Therefore when $r = n$ there is no need to verify condition (2), and the proof given here of Ryser's theorem coincides with M. Hall's proof of his easier theorem.

4. Cruse's theorem

A. Cruse [1] proved the following theorem.

Theorem 3. Let H be an $r \times r$ symmetric latin rectangle on $1, \ldots, n$. Then H can be embedded in the top left hand corner of a symmetric latin square K of order n on $1, \ldots, n$ if and only if the following two conditions are obeyed.

(1) $P(i) \geq 2r - n$ for each $i \in \{1, \ldots, n\}$.

(2) $P(i) \equiv n \pmod 2$ for at least r different $i \in \{1, \ldots, n\}$.

(As in Theorem 2, P(i) is the number of times i occurs in H.)

Proof. 1. Necessity. The necessity of (1) follows from Theorem 2. Since K is symmetric, each $i \in \{1, \ldots, n\}$ will occur an even number of times off the main diagonal. Therefore the number of times it occurs on the main diagonal is $\equiv n \pmod 2$. But K has only $n - r$ diagonal cells which are not in H. Therefore for at most $n - r$ values of $i \in \{1, \ldots, n\}$ is it possible for $P(i) \not\equiv n \pmod 2$. Condition (2) now follows.

2. Sufficiency. For $\nu \in \{1, \ldots, r\}$ let A_ν be the set of integers from $\{1, \ldots, n\}$ which do not occur in the νth row of H. Let $M = \{m_1, \ldots, m_l\}$ be the set of $i \in \{1, \ldots, n\}$ which occur either $2r - n$ times or $2r - n + 1$ times in H (note that the definition of M, the set of marginal elements, is different from the definition of M in Theorem 2). Let C_{r+1} be the set of those $i \in \{1, \ldots, n\}$ for which $P(i) \not\equiv n \pmod 2$. Then, by condition (2), $|C_{r+1}| \leq n - r$. If $m \in M$ and $P(m) = 2r - n + 1$ then $P(m) \not\equiv n \pmod 2$, so that $m \in C_{r+1}$, whereas if $P(m) = 2r - n$ then $P(m) \equiv n \pmod 2$ and so $m \notin C_{r+1}$. It is easy to see now that each $m \in M$ is in exactly $n - r$ of $A_1, \ldots, A_r$, C_{r+1}. If $x \geq 1$ then it is not possible for $r + x$ elements to occur $2r - n$ times in H and the remaining $n - r - x$ elements to occur $\leq r$ times since

$$(r + x)(2r - n) + (n - r - x)r = r^2 + x(r - n) < r^2.$$

Therefore the maximum number of m for which $P(m) = 2r - n$ is r. Therefore we may choose A_{r+1} so that $C_{r+1} \subseteq A_{r+1}$, $|A_{r+1}| = n - r$ and no m for which $P(m) = 2r - n$ is in A_{r+1}. Then for each $\nu \in \{1, \ldots, r+1\}$, $|A_\nu| = n - r$, and also each marginal element is in exactly $n - r$ of $A_1, \ldots, A_{r+1}$. Let $N = \{1, \ldots, r+1\}$ and $\mathfrak{A} = (A_\nu : \nu \in N)$.

We need to show that $\mathfrak{A}$ has a transversal which includes M. For suppose that $\{a_1, \ldots, a_{r+1}\}$ is such a transversal of $\mathfrak{A}$ with $a_\nu \in A_\nu$ for each $\nu \in N$. Then we may form a symmetric $(r+1) \times (r+1)$ latin rectangle $H^\times$ on $1, \ldots, n$ which contains H in its top left hand corner, by putting a_ν in positions $(\nu, r+1)$ and $(r+1, \nu)$ for each $\nu \in N$. Then $H^\times$ obeys conditions (1) and (2) with r replaced by $r + 1$.

To show that $\mathfrak{A}$ has a transversal which includes M we must show that conditions (1) and (2) of Theorem 1 are obeyed. The arguments for these are the same as in the proof of Ryser's theorem (with s replaced by r). The only modification is to notice that the maximum number of times i can appear amongst $A_1, \ldots, A_r$ is $r-(2r-n)=n-r$. If i appears in $n-r$ of $A_1, \ldots, A_r$ then $P(i)=2r-n$, so that $P(i) \equiv n \pmod 2$, so $i \notin A_{r+1}$. Therefore the maximum number of times i can appear amongst $A_1, \ldots, A_{r+1}$ (and therefore in $\underset{\sim}{A}$) is $n-r$.

Theorem 3 now follows by induction on r.

References

1. A. Cruse. On embedding incomplete symmetric latin squares. J. Comb. Theory (to appear).
2. T. Evans. Embedding incomplete latin squares. Amer. Math. Monthly, 67 (1960), 958-61.
3. M. Hall Jr. An existence theorem for latin squares. Bull. Amer. Math. Soc., (1945), 387-8.
4. P. Hall. On representatives of subsets. J. London Math. Soc., 10 (1935), 26-30.
5. A. J. W. Hilton. Embedding an incomplete diagonal latin square in a complete diagonal latin square. J. Comb. Theory A, 15 (1973), 121-8.
6. A. J. Hoffman and H. W. Kuhn. Systems of distinct representatives and linear programming. Amer. Math. Monthly, 63 (1956), 455-60.
7. C. C. Lindner. A survey of finite embedding theorems for partial latin squares and quasigroups. Capital Conference in Graph Theory, Combinatorics and their applications, The George Washington University, 1973.
8. H. B. Mann and H. J. Ryser. Systems of distinct representatives. Amer. Math. Monthly, 60 (1953), 397-401.
9. L. Mirsky. Hall's criterion as a 'self-refining' result. Monatsh. für Math., 73 (1969), 139-46.
10. L. Mirsky and others. Some aspects of transversal theory. Seminar, University of Sheffield, 1967.
11. H. Perfect. Marginal elements in transversal theory. Studies in Pure Mathematics (ed. L. Mirsky), Academic Press, 1971.
12. H. J. Ryser. A combinatorial theorem with an application to latin rectangles. Proc. Amer. Math. Soc., 2 (1951), 550-2.
13. H. J. Ryser. Combinatorial mathematics. The Carus Math. Monographs, No. XIV, Math. Assoc. Amer. (1963).

University of Reading
Reading, England

SOME RESULTS IN SEMI-STABLE GRAPHS*

D. A. HOLTON

Abstract

A characterisation of semi-stable graphs is given in terms of fixed blocks of the automorphism group of a specified subgraph. This characterisation is used to prove that regular graphs are semi-stable, and to indicate that the product of two trees is also semi-stable.

1. Introduction

Throughout all graphs G are on a finite vertex set V, have no loops, multiple edges or directed edges. Graph theoretical concepts are to be found in [1] as are most group theoretical concepts. The remaining permutation group ideas are in [6].

A graph G is <u>semi-stable</u> ([3]) if there exists a vertex v in V such that $\Gamma(G)_v = \Gamma(G_v)$. Here $\Gamma(G)$ is the automorphism group of G, $\Gamma(G)_v$ the subgroup of $\Gamma(G)$ which fixes v and considered as acting on $V\setminus\{v\}$, and G_v is the subgraph of G obtained by removing the vertex v and all edges adjacent to it.

Graphs such as K_n, the complete graph on n vertices, and C_n, the cycle on n vertices, are semi-stable. The smallest graph which is not semi-stable is P_4, the path on 4 vertices.

Robertson and Zimmer [5] have shown that all trees are semi-stable at an end vertex, except the paths P_n $(n \geq 3)$, E_7 (the smallest identity tree - on 7 vertices), and the tree of Figure 1.

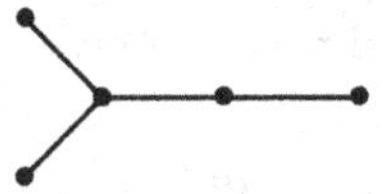

Figure 1

* This work was supported, in part, by A. V. Jennings Industries (Australia) Ltd.

Independently, Heffernan [2] also proved the same result, but in addition showed that P_3 and the graph of Figure 1, are semi-stable at their vertices of degree 2.

In what follows we give a characterisation of semi-stable graphs, and use it, not only to prove that all regular graphs are semi-stable, but also to extend the results on trees to the fact that the product of two arbitrary trees is semi-stable.

2. The characterisation

A fixed block, Δ, of the permutation group Γ acting on the set Ω, is a subset of Ω, such that $\Delta^g = \{\delta^g : \delta \in \Delta, g \in \Gamma\} = \Delta$, for all $g \in \Gamma$ (see [6]).

Theorem 1. G is semi-stable at v if and only if $\Delta = \{w : w \sim v \text{ in } G\}$ is a fixed block of $\Gamma(G_v)$.

Proof. Let G be semi-stable at v. If $w \sim v$ (w adjacent to v) in G, then $w^g \sim v^g$ for all $g \in \Gamma(G)$. If further $g \in \Gamma(G)_v$, then $w^g \sim v$. But since G is semi-stable at v, $\Gamma(G)_v = \Gamma(G_v)$ and so in $V \setminus \{v\}$, $\{w^g : g \in \Gamma(G_v), w \in \Delta\} = \Delta$. Hence Δ is a fixed block of $\Gamma(G_v)$.

On the other hand, let Δ be a fixed block of $\Gamma(H)$; we form G from H by adjoining v to each vertex of Δ. Let $h \in \Gamma(H)$ and let $h' = h(v)$ be the corresponding permutation on V which fixes v. If $w_1 \sim w_2$ in H, then they are also adjacent in G. Then $w_1^h \sim w_2^h$ in H and so $w_1^{h'} \sim w_2^{h'}$ in G. Similarly for $w_1 \not\sim w_2$.

Now suppose $w \in \Delta$. So $v \sim w$ in G. For all $g \in \Gamma(H)$, $w^g \in \Delta$, so $w^g \sim v$ in G. In particular $w^h \sim v$ in G. Hence $w^{h'} = w^h \sim v = v^{h'}$ in G. Similarly if $w \not\sim v$ in G, $w^h \not\sim v$ and so $w^{h'} \not\sim v$ in G. Hence $h' \in \Gamma(G)$ and since h' fixes v, $h' \in \Gamma(G)_v$. So $\Gamma(H) \leq \Gamma(G)_v$. But by [3] Theorem 2, $\Gamma(G)_v \leq \Gamma(G_v)$ for all graphs. Hence in this case G is semi-stable at v, since $H = G_v$ here.

Corollary. If G is regular then G is semi-stable at any vertex.

Proof. Let $\Delta = \{w : w \sim v \text{ in } G\}$. If v is of degree r then all the vertices of Δ in G_v are of degree $r - 1$ and so form a fixed block.

Clearly this characterisation of semi-stability is easier to use in practice than the original definition. One needs no longer to know $\Gamma(G)$, $\Gamma(G)_v$ and $\Gamma(G_v)$ fully. It is sufficient to know a fixed block of $\Gamma(G_v)$, which, as we see from the corollary above, can be found with very little knowledge of $\Gamma(G_v)$, and no knowledge of $\Gamma(G)$.

3. Products of trees

We now exploit the characterisation of semi-stability to prove that the product of any two trees is semi-stable. In particular cases we find a number of vertices at which the product is semi-stable. We first flex our muscles on products of paths.

Theorem 2. For $m \geq 3$, $n \geq 4$ and $m = n = 2$, $P_m \times P_n$ is semi-stable at any vertex.

$P_3 \times P_3$ is semi-stable at all but the vertex of degree 4.

$P_2 \times P_n$, $n \geq 3$ is semi-stable except at vertices of degree 3 whose minimum distance from a vertex of degree 2 is 2.

Proof. We will give a 'proof by example' for a couple of vertices in $P_3 \times P_4$. The complete proof follows in a similar way and can be found in [4]. $P = P_3 \times P_4$ is shown in Figure 2. We first show that v_1 is a vertex of semi-stability

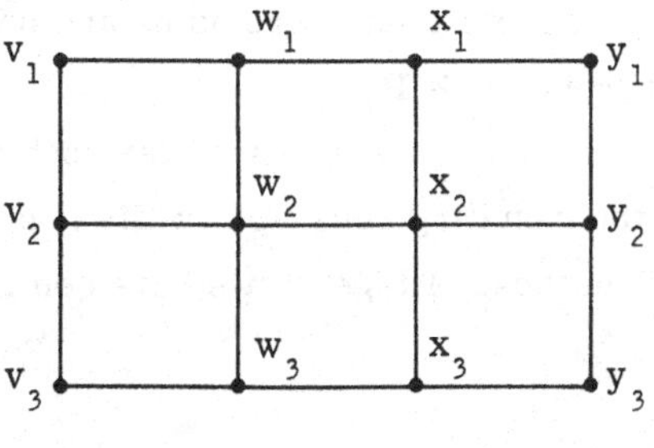

Figure 2

In P_{v_1} the vertices of degree 2 are v_2, v_3, w_1, y_1, and y_3. Now w_2 is the only vertex of degree 4 adjacent to 2 vertices of degree 2, so w_2 must be fixed by $\Gamma(P_{v_1})$. Hence $\{v_2, w_1\}$ is a fixed block of

$\Gamma(P_{v_1})$ since these are the only vertices of degree 2 adjacent to the fixed vertex w_2. Then since v_1 is adjacent to the fixed block $\{v_2, w_1\}$ it must be a vertex of semi-stability of P.

If we consider P_{w_1}, we find that the only vertex of degree 1 is v_1, and so this must be fixed by $\Gamma(P_{w_1})$. Further w_2 is the only vertex of degree 3 at a distance 1 from the fixed vertex v_1, so w_2 itself must be fixed. Finally x_1 is fixed, being the only vertex of degree 2 at a distance 4 from a vertex of degree 1 (v_1) and at a distance 2 from a vertex of degree 3 (w_2). Hence $\{v_1, w_2, x_1\}$ is a fixed block of $\Gamma(P_{w_2})$ and so P is semi-stable at w_2.

With $P_3 \times P_3$ (Figure 3), $\Gamma(P_3 \times P_3) = [S_2]^{S_2}$, and so $|\Gamma(P_3 \times P_3)| = 8$. But $(P_3 \times P_3)_v = C_8$ and hence $|\Gamma((P_3 \times P_3)_v)| = 16$ which cannot possibly equal $|\Gamma(P_3 \times P_3)_v|$.

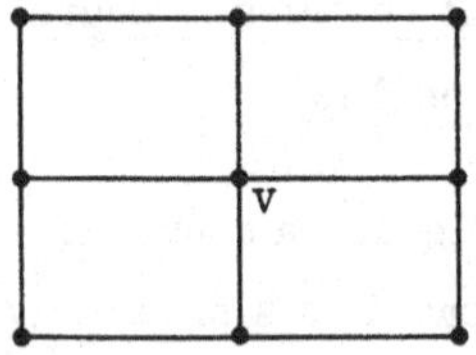

Figure 3

However this is the only case in which extra automorphisms are introduced by the removal of a vertex from $P_3 \times P_3$.

For $P_2 \times P_n$ in Figure 4, trouble comes at a vertex such as v, since $(w_1 w_2)$ is in the group of the resulting subgraph, while it is not in $\Gamma(P_2 \times P_n)$. At vertices other than those of type v we have semi-stability.

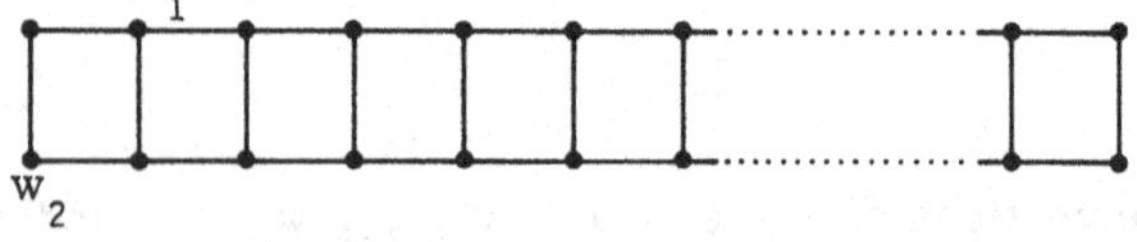

Figure 4

Corollary 1. $P_{m_1} \times P_{m_2} \times \ldots \times P_{m_r}$ *is semi-stable at a vertex of degree* r.

As a result of this corollary we can produce a whole family of identity graphs.

Corollary 2. *For* $m_i \neq m_j$, $i, j = 1, 2, 3, \ldots, r$, $P_v = (P_{m_1} \times P_{m_2} \times \ldots \times P_{m_r})_v$ *is an identity graph, where* v *is a vertex of degree* r *in* P.

Proof. By the orbit-stabiliser relation

$$|\Gamma(P)| = |v^{\Gamma(P)}| . |\Gamma(P)_v| .$$

But if $m_i \neq m_j$ then $|\Gamma(P)| = 2^r$, and $|v^{\Gamma(P)}| = 2^r$, since the vertices of degree r form an orbit of $\Gamma(P)$. Hence

$$1 = |\Gamma(P)_v| = |\Gamma(P_v)|$$

and P_v is an identity graph.

Finally then

Theorem 3. *For trees* T_1, T_2, $T_1 \times T_2$ *is semi-stable.*

Proof. There are again a number of vertices of semi-stability. We choose here the vertex v obtained from an end vertex of T_1 and a penultimate vertex of T_2. Again we indicate the proof by example. A complete proof can be found in [4].

Consider the graph $T = T_1 \times T_2$ of Figure 5. We need to show

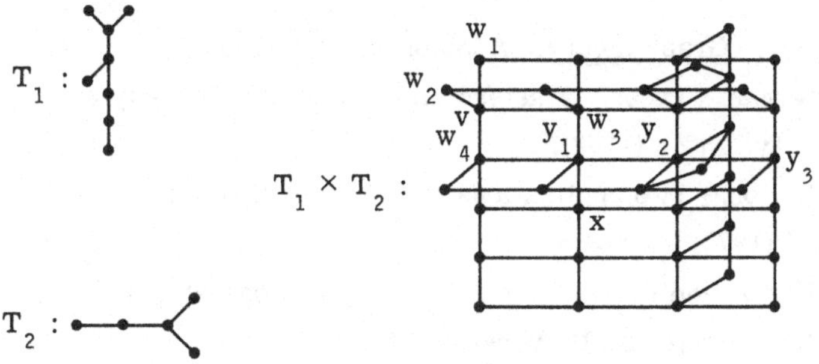

Figure 5

that $\{w_1, w_2, w_3, w_4\}$ is a fixed block of $\Gamma(T_v)$. Since w_1 and w_2 are the only vertices of degree 1 in T_v, $\{w_1, w_2\}$ is a fixed block of $\Gamma(T_v)$. Similarly w_3 is the only vertex of degree 5 a distance 2 from w_1 and w_2, so it too is fixed by $\Gamma(T_v)$.

Now w_4 must go to a vertex of degree 4 which is at distance 2 from w_3, and a distance 4 from w_1 and w_2. The only candidate in T_v is x. Assume there exists $g \in \Gamma(T_v) : w_4^g = x$. Now the 'straight' path w_4, y_1, y_2, y_3 through w_4 must be mapped by g into a similar 'straight' path through x. There is no such 'straight' path through x, so w_4 is fixed.

Hence $\{w_1, w_2, w_3, w_4\}$ is a fixed block of $\Gamma(T_v)$ and so T is semi-stable at v.

Remark. The 'straight' path notion plays an important role in the general proof.

Similar results to that of Corollary 1 to Theorem 2 can also be produced.

As the result of this and other work, it now begins to look as if 'almost all' graphs are semi-stable, in the sense that if s_n is the number of semi-stable graphs on n vertices, and g_n is the number of graphs on n vertices, then

$$\lim_{n\to\infty} \frac{s_n}{g_n} = 1.$$

Hence it becomes increasingly more interesting to find all graphs which are <u>not</u> semi-stable. It would be interesting to characterise such graphs.

References

1. F. Harary. Graph theory. Addison-Wesley (1969).
2. P. Heffernan. Trees. Masters dissertation, University of Canterbury, N. Z. (1972).
3. D. A. Holton. Two applications of semi-stability. Discrete Math., 4 (1973), 151-8.
4. D. A. Holton and D. D. Grant. Stability and operations on graphs. Pure Math. Preprint, University of Melbourne.

5. N. Robertson and J. A. Zimmer. Automorphisms of subgraphs obtained by deleting a pendant vertex. J. Comb. Th., 12(B) (1972), 169-73.

6. H. Wielandt. Finite permutation groups. Academic Press (1964).

University of Melbourne,
Parkville, Victoria,
Australia

HEREDITARY PROPERTIES AND P-CHROMATIC NUMBERS

RHYS PRICE JONES

Abstract

We define a topology on the set of all graphs, and characterise the continuous functions. We define P-chromatic numbers, and place bounds on their values when P is an hereditary property. We define a P-chromatic polynomial. We then look briefly at the families D_k of k-degenerate graphs. We state two theorems relating families of the type D_k^i, where D_k^i is the family of graphs whose D_k-chromatic numbers do not exceed i.

1. Introduction

This paper is a brief synopsis of some of the results which I proved in [1]. Graphs shall be finite and undirected with no loops or multiple edges. The notation $H \leq G$ means that H is a subgraph of G. $H < G$ means that H is a vertex-induced subgraph of G. We do not distinguish between isomorphic graphs. U shall denote the set of all graphs. Definitions not given here can be found in [2].

2. A topology

Let P be any family of graphs. P shall also denote the property that a graph be a member of the family P. We shall use the terms 'family' and 'property' interchangeably. P is <u>weakly hereditary</u> if, whenever $G \in P$ and $H \leq G$, then $H \in P$. We let T denote the family of all weakly hereditary properties. One more definition: A function $f : U \to U$ is <u>monotonic</u> if, whenever $H \leq G$, then $f(H) \leq f(G)$.

Theorem 2.1. (U, T) <u>is a topological space whose open sets are the members of</u> T. <u>A function</u> $f : (U, T) \to (U, T)$ <u>is continuous if and only if it is monotonic.</u> //

In [1], we indicate how this theorem may be used to form new hereditary properties from old. No other applications of the theorem have been found.

3. More definitions

Let P be any property of graphs and let G be a graph. The subset $S \subset V(G)$ is a P-set if $\langle S \rangle$, the subgraph of G induced by the vertices of S, has property P. For convenience, the empty set $\emptyset$ is always to be regarded as a P-set. An (m-P)-colouring of G is a partition of V(G) into m P-sets. P^m denotes the family of all graphs which admit an (m-P)-colouring. The P-chromatic number of G, denoted by $\chi_P(G)$, is the smallest integer m for which $G \in P^m$. Notice that $\chi_{D_0}(G)$, where D_0 denotes the property that a graph be totally disconnected, is the usual chromatic number of G.

A weakly hereditary family P is hereditary if $P \neq \emptyset$ and $P \neq U$. For any hereditary property P there is a non-negative integer k(P), called the completeness of P, so that the complete graph $K_{k(P)+1} \in P$ but $K_{k(P)+2} \notin P$.

The P-stability number of G, denoted by $M_P(G)$, is the largest number of vertices of G which can form a P-set.

4. Bounds on the P-chromatic numbers

Theorem 4.1. Let P be an hereditary property and let G be a graph with p vertices. Let $M_P(G) = M$ and let $k(P) = k$. Then:

$$\left\{\frac{p}{M}\right\} \leq \chi_P(G) \leq \left\{\frac{p-M}{k+1}\right\} + 1,$$

where $\{r\}$ denotes the least integer not less than r. //

Corollary 4.2. $\chi_P(K_p) = \left\{\frac{p}{k+1}\right\}$. //

In [1], we prove a partial generalisation of the Nordhaus-Gaddum Theorem. A family P of graphs is m-colourable, if every member of the family can be m-coloured (in the traditional sense). The property P is m-like if $K_p \in P$ for every $1 \leq p \leq m+1$, and whenever S is a P-set and a vertex $v \notin S$ is adjacent to at most m vertices of S, then $S \cup \{v\}$ is also a P-set. $\bar{G}$ will denote the complement of the graph G.

Theorem 4.3. Let G be a graph with p vertices. Let P be a property.

(i) If P is m-colourable and $m \geq 1$, then

$$\chi_P(G) \cdot \chi_P(\bar{G}) \geq \frac{p}{m^2} \text{ and } \chi_P(G) + \chi_P(\bar{G}) \geq \frac{2\sqrt{p}}{m}.$$

(ii) If P is m-like and $m = 0$ or 1, then

$$\chi_P(G) + \chi_P(\bar{G}) \leq \frac{p+2m+1}{m+1} \text{ and } \chi_P(G) \cdot \chi_P(\bar{G}) \leq \frac{1}{4}\left(\frac{p+2m+1}{m+1}\right)^2. \quad //$$

We believe that part (ii) of Theorem 4.3 can be extended to all non-negative integers m.

5. A P-chromatic polynomial

For any family P of graphs, and for any graph G, we let $f_P(G, t)$ be the number of (t-P)-colourings of G. We impose conventions similar to those imposed by Ronald Read in [5]. (These can be summarised by saying that we insist that the vertices of G shall be labelled, and that the 'colours' - i.e. the parts of the partition - shall be labelled.)

The following theorem is proved in [1] using the Principle of Inclusion and Exclusion. The statement of the theorem is so sweeping as to make us doubt its usefulness:

Theorem 5.1. For any family P of graphs, and for any graph G, $f_P(G, t)$ is a polynomial in t. //

6. New families from old

Recall that P^m is the family of graphs whose P-chromatic number does not exceed m. k(P) is the completeness of the hereditary family P. We shall state two theorems which indicate what happens to the completeness and chromatic numbers when new families are derived from old.

Theorem 6.1. Let P be an hereditary property and let G be a graph:

(i) P^m is also hereditary.

(ii) $k(P^m) = m(k(P) + 1) - 1$.

(iii) $\chi_{(P^m)}(G) = \{\frac{1}{m} \chi_P(G)\}$.

Proof. We shall prove part (iii).

Firstly, let $\chi_{(P^m)}(G) = n$. Then we have a partition $V(G) = U_1 \cup U_2 \cup \ldots \cup U_n$ where each U_i is a P^m-set. Accordingly $U_i = V_{i1} \cup V_{i2} \cup \ldots \cup V_{im}$ where each V_{ij} is a P-set. The subsets V_{ij} $(1 \leqq i \leqq n;\ 1 \leqq j \leqq m)$ are the parts of a partition on $V(G)$ into mn P-sets. This proves that $\chi_P(G) \leqq m.\ \chi_{(P^m)}(G)$.

Now let $V(G)$ be partitioned into $\chi_P(G)$ P-sets. The union of any collection of m or less of these is a P^m-set. And we can form $\{\frac{1}{m} \chi_P(G)\}$ of these whose union is the whole of $V(G)$. This proves that $\chi_{(P^m)}(G) \leqq \{\frac{1}{m} \chi_P(G)\}$. //

Theorem 6.2. Let P and Q be hereditary properties and let G be a graph:

(i) $P \cup Q$ and $P \cap Q$ are also hereditary.

(ii) $k(P \cup Q) = \text{Max}\{k(P),\ k(Q)\}$ and $k(P \cap Q) = \text{Min}\{k(P), k(Q)\}$

(iii) $\chi_{P \cup Q}(G) \leqq \text{Min}\{\chi_P(G),\ \chi_Q(G)\}$ and $\chi_{P \cap Q}(G) \geqq \text{Max}\{\chi_P(G),\ \chi_Q(G)\}$. //

Our next step is to find relationships between families of the forms P^i and Q^j for various families P and Q and integers i and j. Here we find that we must depart from our general theory and look at a more restricted class of families.

7. k-degenerate graphs

Let $\delta(H)$ denote the minimum vertex-degree of a graph H. The strength of a graph G, denoted by $\sigma(G)$, is the largest of the minimum degrees of its subgraphs. Symbolically:

$$\sigma(G) = \text{Max}\{\delta(H) : H \leq G\} = \text{Max}\{\delta(H) : H < G\}.$$

A graph is k-degenerate if its strength does not exceed k. The family of all k-degenerate graph is denoted by D_k. The families of k-degenerate graphs have been studied by Lick and White [4]. For convenience, we denote the D_k-chromatic number of a graph G by $\rho_k(G)$.

Notice that D_0 is the family of totally disconnected graphs, so that $\rho_0(G)$ is the (ordinary) chromatic number of G. D_1 is the family of acyclic graphs (or forests), so that $\rho_1(G)$ is the point arboricity of G.

D_k is an hereditary family of completeness k. D_k is (k+1)-colourable and is k-like.

It is known that the largest eigenvalue $\lambda(G)$ of the adjacency matrix of a graph G has the following properties:

$\lambda 1$: If $H \leq G$, then $\lambda(H) \leq \lambda(G)$.

$\lambda 2$: $\lambda(G) \geq \sigma(G)$.

In [6]. Szekeres and Wilf prove that $\rho_0(G) \leq 1 + \mu(G)$, where μ is any real valued function defined on U which has properties $\lambda 1$ and $\lambda 2$. In [1], we generalise this result:

Theorem 7.1. Let μ be any function which maps every graph onto a real number, and which satisfies conditions $\lambda 1$ and $\lambda 2$. Then $\rho_k(G) \leq 1 + [\frac{\mu(G)}{k+1}]$ where [r] denotes the largest integer not exceeding r. //

Corollary 7.2.

(i) $\rho_k(G) \leq 1 + [\frac{\sigma(G)}{k+1}]$,

(ii) $\rho_k(G) \leq 1 + [\frac{\rho_0(G)}{k+1}]$,

(iii) $\rho_k(G) \leq 1 + [\frac{\lambda(G)}{k+1}]$,

(iv) $\rho_k(G) \leq 1 + [\frac{\Delta(G)}{k+1}]$ where $\Delta(G)$ is the maximum vertex degree of G. //

Part (i) was proved in [4] and (iii) was proved in [3]. All of these results are best possible in that the bounds can be attained; in fact they are all attained whenever G is a complete graph. Brooks' Theorem characterises those graphs for which the bound of (iv) is attained in the case $k = 0$. It seems to be very difficult to find such a characterisation when $k > 0$.

The results of parts (i) and (ii) of Corollary 7.2 are used in [1] to prove the next two theorems:

Theorem 7. 3. For any integers $k \geq 0$ and $m \geq 1$:

$$D_0^{m(k+1)-1} \subset D_k^m \subset D_0^{m(k+1)} . \quad //$$

Theorem 7. 4. For any positive integers m and k:

$D_{mk-1} \subset D_{k-1}^m$. //

Conclusion

Many familiar colouring theorems are special cases of more general results. It seems likely that we shall gain greater insight into graph-colouring theory by considering the more general situation.

References

1. R. P. Jones. Partition numbers of graphs. M. Sc. Thesis, Univ. of Calgary, May 1973.
2. F. Harary. Graph theory. Addison-Wesley, Reading, Mass. (1969).
3. D. R. Lick. A class of point partition numbers. Lecture notes in Mathematics 186 - Recent trends in graph theory, Springer-Verlag, Berlin, Heidelburg, New York (1971), pp. 185-90.
4. D. R. Lick and A. T. White. k-Degenerate graphs. Canad. J. Math., XXII (1970), 1082-95.
5. R. C. Read. An introduction to chromatic polynomials. J. Combinatorial Theory, 4 (1968), 52-71.
6. G. Szekeres and H. S. Wilf. An inequality for the chromatic number of a graph. J. Combinatorial Theory, 4 (1968), 1-3.

Royal Holloway College,
London, England

SOME PROBLEMS CONCERNING COMPLETE LATIN SQUARES

A. D. KEEDWELL

The concept of row complete latin squares arises in statistics in connection with the design of sequential experiments. The earliest papers known to the author which discuss the subject from this point of view are B. R. Bugelski [1] and E. J. Williams [10].

A latin square L is called row complete if each ordered pair α, β of its elements occur as adjacent elements once (and necessarily only once) in some row of the square.

We may similarly define a column complete latin square. If a latin square is both row complete and column complete, it is called complete.

All known examples of row complete latin squares can be made column complete as well by suitably rearranging their rows. Thus, for example, the first latin square shown in Fig. 1 is row complete but not column complete. It becomes column complete when the last two rows are interchanged as in the second square shown in Fig. 1. This observation

0	1	2	3
1	3	0	2
3	2	1	0
2	0	3	1

0	1	2	3
1	3	0	2
2	0	3	1
3	2	1	0

Figure 1

leads us to our first question: namely, 'Do there exist row complete latin squares which cannot be made column complete as well by suitably reordering their rows?' (See note 1 on page 96.)

It is easy to show that row complete latin squares of every even order $n = 2m$ exist. We have the following theorem, first proved by E. J. Williams [10].

Theorem 1. We may obtain a row complete latin square L of any even order 2m by taking as first row the integers 0, 1, 2m-1, 2, 2m-2, 3, 2m-3, ..., m+1, m and forming each subsequent row by adding 1 modulo 2m to the elements of the preceding row.

The success of the construction is consequent upon the fact that the differences between adjacent elements of the prescribed first row are all different when taken modulo 2m. The square L which is obtained is a Cayley addition table of the cyclic group of order 2m. It may be made column complete as well as row complete by rearranging its rows in such a way that the first column becomes the same as the first row.

The situation with regard to row complete latin squares which can be constructed by means of abelian groups has been completely resolved by B. Gordon [4], by means of theorems 2 and 3 below:

Theorem 2. A sufficient condition for the existence of a complete latin square L of order n is that there exist a finite group G of order n with the property that its elements can be arranged in a sequence $a_1, a_2, \ldots, a_n$ in such a way that the partial products $b_1 = a_1$, $b_2 = a_1a_2$, $b_3 = a_1a_2a_3$, ..., $b_n = a_1a_2 \ldots a_n$ are all distinct.

A group G which satisfies the conditions of theorem 2 is called sequenceable.

Theorem 3. A finite abelian group G is sequenceable if and only if it is the direct product, $G = A \times B$, of two groups A and B such that A is cyclic of order 2^k, $k > 0$, and B is of odd order.

It follows from theorem 3 that sequenceable abelian groups of odd order do not exist. However, the existence of non-abelian sequenceable groups of odd order is not ruled out. In particular, N. S. Mendelsohn [6] has shown that the non-abelian group of order 21 is sequenceable and the present author conjectures that in fact every non-abelian group of odd order with two generators is sequenceable. In support of this conjecture he looked for, and found, a sequencing for the non-abelian group of order 27 on two generators a few hours before the present paper was communicated to the conference. (The sequencing is given at the end of the paper.) By virtue of theorem 2, the existence of these sequencings ensures the existence of row complete latin squares of orders 21 and 27. Up to the present, these are the only odd orders for which complete latin squares have been obtained.

As regards non-abelian groups of even order, J. Dénes and E. Török [3] have shown that the dihedral groups D_5, D_6, D_7 and D_8 of orders 10, 12, 14 and 16 respectively are sequenceable but that there are no other sequenceable non-abelian groups of orders less than or equal to 14. The non-sequenceability of the dihedral groups D_3 and D_4 were already known earlier. (See B. Gordon [4].)

So far, no row complete latin squares other than those based on groups have been found and it is known that up to order 6 inclusive no others exist. (See D. Warwick [9].)

The present author would like to pose the following questions:

(i) If a square matrix is row complete, column complete and row-wise latin (that is, each element occurs exactly once in each row) is it necessarily column-wise latin as well? More generally, do any three of the constraints row complete, column complete, row-wise latin and column-wise latin when imposed on a square matrix together imply the fourth? (See the question posed earlier in this paper.)

(ii) Do there exist row complete latin squares which do not satisfy the quadrangle criterion? (In other words, do there exist quasigroups which are not isotopic to any group but whose Cayley multiplication tables represent row complete latin squares?)

(iii) Do there exist row complete latin squares which have column complete orthogonal mates?

(iv) Do there exist complete latin squares which have complete orthogonal mates?

In connection with problems (iii) and (iv), we should like to point out that a necessary condition (which is also sufficient if G is abelian or soluble) for a group G to have a complete mapping is that there exist an ordering $a_1, a_2, \ldots, a_n$ of its elements such that the product $a_1 a_2 \ldots a_n$ is equal to the identity element of G. (See L. J. Paige [7] and [8] and M. Hall and L. J. Paige [5].) Since the existence of n complete mappings (n disjoint transversals) is necessary for a latin square of order n representing the multiplication table of G to have an orthogonal mate and since, for an abelian group of order n with a unique element g of order two, the product $a_1 a_2 \ldots a_n$ of the n distinct elements is equal to g (L. J. Paige [7]), a complete latin square based on a sequenceable abelian group cannot have an orthogonal mate.

Row complete latin squares and latin rectangles have connections with graph theory. In particular, the former have been used to give a partial answer to the question 'For which integers n does the complete directed graph on n vertices have a decomposition into n disjoint Hamiltonian paths?' N. S. Mendelsohn [6], and also J. Dénes and E. Török [3], have pointed out that if n is an integer for which a row complete latin square of order n exists, then the required decomposition is always possible. If the vertices of the graph are labelled by the same set as the elements of the row complete latin square, the rows of the latter prescribe the Hamiltonian paths of the required decomposition. Consequently, decompositions of the required kind certainly exist when n is even and when n is equal to 21 or 27.

The present author has investigated the analogous question for complete undirected graphs with n vertices. If such a graph has a decomposition into disjoint Hamiltonian paths, then the number of paths is n/2. Consequently, the question is only meaningful if the number n of vertices is even. In that case, the answer is 'Yes, always'. However, we have been able to prove more. (But see note 2 on page 96.)

A latin rectangle of m rows and 2m columns is called row <u>complete</u> if each <u>un</u>ordered pair of its elements occur as adjacent elements once (and necessarily only once) in some row of the rectangle. We have the following theorem.

Theorem 4. <u>Row complete</u> m × 2m <u>latin rectangles exist for every positive integer</u> m <u>and, correspondingly, (i) for every positive integer</u> m, <u>the complete undirected graph on</u> 2m <u>vertices has a decomposition into</u> m <u>disjoint Hamiltonian paths. Also (ii) the complete undirected graph on</u> 2m + 1 <u>vertices has a decomposition into</u> m <u>disjoint circuits each of length</u> 2m + 1. <u>Moreover, (iii) every complete undirected graph on an odd number</u> 2m + 1 <u>of vertices has an Eulerian line with the property that, when a certain vertex and all the edges through</u>

it are deleted, the remaining portions of the Eulerian line are Hamiltonian paths of the residual graph on 2m vertices.

Proof. Consider the row complete latin square L formed by taking as first row the integers

0, 1, 2m-1, 2, 2m-2, ..., k, 2m-k, ..., m+1, m

(as in theorem 1) and whose subsequent rows are obtained, each from its predecessor, by adding 1 modulo 2m to its elements (as in Fig. 2).

The last m rows are the same as the first but in reverse order. Thus, in the latin rectangle formed by the first m rows every unordered pair of elements occurs as a consecutive pair just once in some row. The first m rows define the required Hamiltonian decomposition of the undirected graph on 2m vertices. This proves (i).

0	1	2m-1	2	2m-2	...	m+2	m-1	m+1	m
1	2	0	3	2m-1	...	m+3	m	m+2	m+1
.	.	.	.	.	...	.	.	.	.
m-1	m	m-2	m+1	m-3	...	1	2m-2	0	2m-1
m	m+1	m-1	m+2	m-2	...	2	2m-1	1	0
.	.	.	.	.	...	.	.	.	.
2m-1	0	2m-2	1	2m-3	...	m+1	m-2	m	m-1

Figure 2

Now let an additional column be adjoined to the latin rectangle formed by the first m rows of L, as shown in Fig. 3. Since the leading entries of these m rows together with their final entries exactly cover the set 0, 1, ..., 2m-1, the m augmented rows may be read from left to right in consecutive order, the first being successor to the last, (as shown in Fig. 3) to define an Eulerian circuit of the complete undirected graph on 2m + 1 vertices which has the properties described in statement (iii) of the theorem. Finally, to obtain the disjoint cycle decomposition whose existence is claimed in statement (ii), it is only necessary to regard each separate row of the augmented rectangle as defining one such cycle when the row in question is itself read cyclically, the first of its entries being regarded as successor to the last.

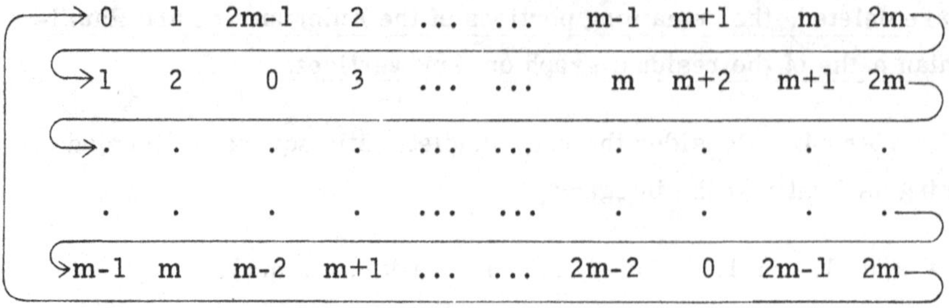
0 1 2m-1 2 ... ... m-1 m+1 m 2m
1 2 0 3 ... ... m m+2 m+1 2m
. . . .
. . . .
m-1 m m-2 m+1 ... ... 2m-2 0 2m-1 2m

Figure 3

It would be interesting to investigate for what other undirected graphs property (iii) of theorem 4 above is valid.

A more detailed account of complete latin squares which includes proofs of all the results quoted above will be found in a forthcoming book, J. Dénes and A. D. Keedwell [2]. The book also discusses a number of other connections between graph theory and various kinds of latin square.

References

1. B. R. Bugelski. A note on Grant's discussion of the latin square principle in the design of experiments. Psychological Bull., 46 (1949), 49-50.

2. J. Dénes and A. D. Keedwell. Latin squares and their applications. Akadémiai Kiadó, Budapest and English Universities Press, London, and Academic Press, New York (1974).

3. J. Dénes and É. Török. Groups and graphs. Combinatorial theory and its applications, 257-89. North Holland, Amsterdam (1970).

4. B. Gordon. Sequences in groups with distinct partial products. Pacific J. Math., 11 (1961), 1309-13.

5. M. Hall and L. J. Paige. Complete mappings of finite groups. Pacific J. Math., 5 (1955), 541-9.

6. N. S. Mendelsohn. Hamiltonian decomposition of the complete directed n-graph. Theory of graphs (Proc. Colloq. Tihany, 1966), 237-41. Academic Press, New York (1968).

7. L. J. Paige. A note on finite abelian groups. Bull. Amer. Math. Soc., 53 (1947), 590-3.

8. L. J. Paige. Complete mappings of finite groups. Pacific J. Math., 1 (1951), 111-6.

9. D. Warwick. Methods of construction and a computer search for row complete latin squares. Undergraduate special study, University of Surrey (1973).

10. E. J. Williams. Experimental designs balanced for the estimation of residual effects of treatments. Austr. J. Sci. Research, Ser. A, 2 (1949), 149-68.

Appendix

Sequencing for the group $G = \{a, b \mid a^9 = b^3 = e, ab = ba^4\}$.

The elements $b_1, b_2, \ldots, b_n$ of theorem 2 are shown on the upper line and the elements $a_1, a_2, \ldots, a_n$ on the lower line. Thus,

$$a_1 = e, \; a_2 = b^2a^6, \; a_3 = b^2a^4$$

$$b_1 = e, \; b_2 = e.b^2a^6 = b^2a^6, \; b_3 = e.b^2a^6.b^2a^4 = ba$$

$$e \quad b^2a^6 \quad ba \quad a^7 \quad ba^7 \quad b^2a^8 \quad b^2a^7 \quad a^4 \quad b$$

$$e \quad b^2a^6 \quad b^2a^4 \quad b^2 \quad ba^6 \quad ba^7 \quad a^8 \quad ba^3 \quad ba^2 \quad ba^5$$

$$b^2a^5 \quad a \quad a^2 \quad a^5 \quad ba^2 \quad ba^4 \quad a^3 \quad b^2a^2 \quad b^2 \quad b^2a^4$$

$$ba^8 \quad a \quad a^3 \quad b \quad a^2 \quad b^2a^2 \quad b^2a^8 \quad a^7 \quad a^4 \quad ba$$

$$a^8 \quad b^2a^3 \quad ba^6 \quad b^2a \quad ba^3 \quad ba^8 \quad ba^5 \quad a^6$$

$$b^2a \quad b^2a^3 \quad ba^4 \quad b^2a^5 \quad a^5 \quad a^6 \quad b^2a^7$$

University of Surrey,
Guildford, England

Note 1. Since the above article was written, the author has shown that a sufficient condition that a row complete latin square can be made column complete as well by suitably reordering its rows is that is represents the multiplication table of a group or of an inverse property loop G which satisfies the identity $(gh)(h^{-1}k) = gk$ for all g, h, k in G.

Note 2. The author is indebted to Professor G. A. Dirac for drawing his attention to the fact that the existence of decompositions of the complete undirected graph with n vertices into disjoint Hamiltonian paths when n is even was first shown by Mr. Walecki of Condorcet. For the details, see E. Lucas, Recreations Mathematiques, Vol. II, pages 162-3, Gauthier-Villars, Paris (1883).

NECKLACE ENUMERATION WITH ADJACENCY RESTRICTIONS

E. KEITH LLOYD

1. Introduction

Let G, H be two finite graphs (possibly with loops but with no multiple edges) having vertex sets V(G), V(H) and edge sets E(G), E(H). An edge joining vertices u and v is denoted by [u, v] and, therefore, a loop at u by [u, u]. A function $f : V(G) \to V(H)$ is called a <u>homomorphism</u> if it preserves adjacencies, i. e. if $[u, v] \in E(G)$ implies $[f(u), f(v)] \in E(H)$. The set of all such homomorphisms is denoted by hom(G, H). The set of all one-one homomorphisms from G to itself forms the <u>automorphism group</u> of G denoted by aut(G). It is often desirable to regard $f, g \in$ hom(G, H) as equivalent if there exists $a \in$ aut(G) such that $f = ga$. The set of equivalence classes is denoted by hom(G, H)/aut(G).

A <u>colouring</u> of a graph G is a function $k : V(G) \to C$ where C is a finite set whose elements are called <u>colours.</u> The following general problem is of some interest: how many colourings of G are there subjec to certain specified adjacency restrictions - i. e. as well as C a matrix B is given with $(B)_{ij} = 1$ (or 0) if colour i may (or may not) be adjacent to colour j. Of course, B may be regarded as the adjacency matrix of a graph H. Each colouring of G (subject to the restrictions) corresponds to an element of hom(G, H) and conversely. In the classical colouring problem adjacent vertices may not be assigned the same colour so in this case $(B)_{ij} = 1$ $(i \neq j)$ and $(B)_{ii} = 0$. The corresponding graph H is a complete graph.

One of the problems considered in [1] is the number of monocyclic aromatic compounds formed from three kinds of atoms X, Y, Z. This is equivalent to finding the number of necklaces with three kinds of beads. Explicit formulae were obtained and in [2] the material was expressed in terms of Pólya's theorem and extended to other compounds. Empirical results for small numbers of atoms were also obtained in [1] under the restriction that no two Z atoms be adjacent. This may be viewed as a graph colouring problem with G a circuit graph and H the graph of fig. 1. The solution to this particular problem may be deduced

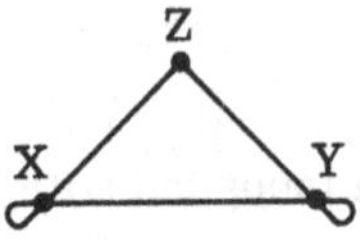

Figure 1

from a more general necklace enumeration problem solved in this paper.

2. Adjacency matrices

It is well known that if B is the adjacency matrix of a graph H then $(B^m)_{ij}$ is the number of paths of length m from the ith vertex of H to the jth vertex. For present purposes it is more convenient to work with a matrix whose entries are names of vertices. Let $V(H) = \{X_1, X_2, \ldots, X_n\}$ and T be the matrix with $(T)_{ij} = X_j$ (or 0) if [i, j] is (or is not) an edge of H. Then $(T^m)_{ij}$ is a formal sum of products of vertices and each such product corresponds to a path of length m in H from X_i to X_j, but since a path of length m contains m + 1 vertices the product omits the first vertex in the path. Using := to denote definitional equality the row vector $\underline{w} := (X_1, X_2, \ldots, X_n)$ and the column vector $\underline{u} :=$ col (1, 1, ..., 1). To find all paths of length m in H it suffices to calculate the product $\underline{w}T^m\underline{u}$, the vector $\underline{w}$ restoring the missing initial vertices. Paths with initial vertex in $W \subseteq V(G)$ and final vertex in $U \subseteq V(G)$ may be generated by replacing $\underline{u}$ by a characteristic vector for U and $\underline{w}$ by a quasi-characteristic vector (with $(\underline{w})_i = X_i$ [or 0] if $X_i \in$ [or $\notin$] W).

The generation of paths in this way appears in the literature in several places in a variety of disguises, see for example Conway [3], Kaufmann [4] and Read [5].

3. Cycle indices

If D is a finite set with a group A acting on it then each $a \in A$ partitions D into a set of disjoint cycles $C(a)$. The cycle index of the action is defined as the polynomial

$$Z(A; s_n) := \frac{1}{|A|} \sum_{a \in A} \prod_{c \in C(a)} s_{r(c)}$$

where $r(c)$ is the length of the cycle c, and each s_n is an indeterminate.

A typical application of Pólya's theorem consists of interpreting s_n as the classical power-sum symmetric function. For example, a generating function for the number of necklaces with m beads made from three types of beads X, Y, Z is obtained by taking $s_n = X^n + Y^n + Z^n$ in the cycle index of the dihedral group. In section 5 the restricted necklace problem is solved by interpreting the products of s_n as products of appropriate matrices.

4. Orbit graphs

Let G, H be two graphs. Each $a \in \mathrm{aut}(G)$ defines a new graph G', called the orbit graph of a, as follows: the vertices of G' are the orbits of $V(G)$ under the action of a. Vertices $[u]$, $[v] \in V(G')$ are adjacent if and only if there are vertices $u' \in [u]$, $v' \in [v]$ with u' and v' adjacent in G.

5. Necklaces

Theorem. Let $\{X_1, \ldots, X_n\}$ be a set of n different colours, T and B the $n \times n$ matrices defined by $(T)_{ij} = X_j$ (or 0) if colour X_j may (or may not) be adjacent to X_i and $(B)_{ij} = 1$ (or 0) if $(T)_{ij} \neq 0$ (or $= 0$). Also let $R(x_1, \ldots, x_n)$ be the number of unrooted, undirected necklaces with x_i beads of colour X_i with permitted adjacencies specified by B. Then for $m \geq 1$

$$(5.1) \qquad \sum_{x_1 + \ldots + x_n = m} R(x_1, \ldots, x_n) X_1^{x_1} \ldots X_n^{x_n}$$

$$= \tfrac{1}{2} \{ \frac{1}{m} \sum_{r|m} \phi(r) \mathrm{tr}(T_r^{m/r}) + \theta_m \}$$

where T_r is the matrix with $(T_r)_{ij} = [(T)_{ij}]^r$,

$$\theta_m = [X_1, \ldots, X_n]T_2^{(m-1)/2}\mathrm{col}[(B)_{11}, \ldots, (B)_{nn}] \text{ (m odd), and}$$

$$\theta_m = \tfrac{1}{2}\{[(T_2)_{11}, \ldots, (T_2)_{nn}]T_2^{(m-2)/2}\mathrm{col}[(B)_{11}, \ldots, (B)_{nn}]$$
$$+ [X_1, \ldots, X_n]T_2^{(m-2)/2}T\ \mathrm{col}[1, \ldots, 1]\} \text{ (m even).}$$

Also,

$$(5.2)\quad \sum_{x_1=0}^{\infty} \cdots \sum_{x_n=0}^{\infty} R(x_1, \ldots, x_n)X_1^{x_1} \cdots X_n^{x_n}$$
$$= 1 - \tfrac{1}{2}\{\sum_{r=1}^{\infty} \frac{\phi(r)}{r} \log\det(I - T_r) + \theta\},$$

where

$$\theta = [X_1, \ldots, X_n](I - T_2)^{-1}\mathrm{col}[(B)_{11}, \ldots, (B)_{nn}]$$
$$+ \tfrac{1}{2}\{[(T_2)_{11}, \ldots, (T_2)_{nn}](I - T_2)^{-1}\mathrm{col}[(B)_{11}, \ldots, (B)_{nn}]$$
$$+ [X_1, \ldots, X_n](I - T_2)^{-1}\mathrm{col}[1, \ldots, 1]\}.$$

Proof. From the viewpoint adopted above the problem is to find $|\mathrm{hom}(G, H)/\mathrm{aut}(G)|$ where G is a circuit of length m (so that $\mathrm{aut}(G)$ is the dihedral group D_m) and H is the graph with adjacency matrix B. The cycle index of the dihedral group is

$$Z(D_m; s_c) = \frac{1}{2m} \sum_{r|m} \phi(r)s_r^{m/r} + \tfrac{1}{2}\theta$$

where

$$\theta = s_1 s_2^{(m-1)/2} \qquad \text{(m odd)}$$

$$\theta = \tfrac{1}{2}[s_2^{m/2} + s_1 s_2^{(m-2)/2} s_1] \qquad \text{(m even).}$$

The orbit graphs corresponding to the various terms are:

$s_r^{m/r}$

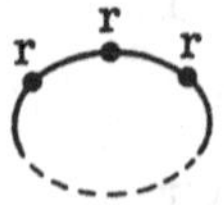

(circuit with m/r vertices)

$s_1 s_2^{(m-1)/2}$

$s_2^{(m-2)/2}$

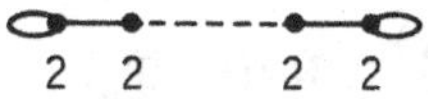

$s_1 s_2^{(m-2)/2} s_1$

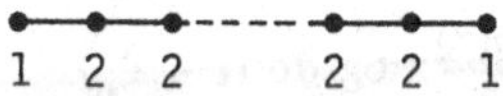

The numbers on the vertices indicate the size of the orbit and the last product has been written in an unusual way to reflect the linear ordering of the vertices in the orbit graph.

The entry $(T^r)_{ii}$ generates rooted dicircuits with X_i as root. To remove the rooting it is only necessary to replace the terms $s_r^{m/r}$ in the cycle index by $T_r^{m/r}$ and to take the trace of the resulting matrix. To remove the orientation of the circuit the appropriate substitutions must be made in the remaining terms of the cycle index. The leftmost factor in each product is replaced by a 'start' vector: $[X_1, \ldots, X_n]$ for the singleton unlooped vertices and $[(T_2)_{11}, \ldots, (T_2)_{nn}]$ for the doubleton looped vertices. The remaining factors s_1, s_2 are replaced by T_1, T_2 respectively. Finally the 'stop' vector is $\text{col}[(B)_{11}, \ldots, (B)_{nn}]$ (or $\text{col}[1, \ldots, 1]$) if there is (or is not) a loop at the right hand end of the path. The presence of a loop means that any bead which may not be adjacent to itself cannot be assigned to the looped orbit.

Together these substitutions yield (5.1). The generating function (5.2) is obtained by summing (5.1) over m. The term θ follows immediately since

$$\sum_{m \text{ odd}} T_2^{(m-1)/2} = \sum_{m \text{ even}} T_2^{(m-2)/2} = (I - T_2)^{-1}.$$

The term 1 in (5.2) corresponds to the empty necklace with $x_1 = \ldots = x_n = 0$. Finally

$$\sum_{m=1}^{\infty} \frac{1}{m} \sum_{r|m} \phi(r) \operatorname{tr} T_r^{m/r}$$

$$= \sum_{r=1}^{\infty} \frac{\phi(r)}{r} \sum_{n=1}^{\infty} \frac{1}{n} \operatorname{tr} T_r^n \qquad (m = nr)$$

$$= \sum_{r=1}^{\infty} \frac{\phi(r)}{r} \operatorname{tr} \sum_{n=1}^{\infty} \frac{1}{n} T_r^n$$

$$= - \sum_{r=1}^{\infty} \frac{\phi(r)}{r} \operatorname{tr} \log(I - T_r)$$

$$= - \sum_{r=1}^{\infty} \frac{\phi(r)}{r} \log \det(I - T_r).$$

6. Acknowledgements

The author has pleasure in thanking Professor A. T. Balaban for helpful correspondence.

References

1. A. T. Balaban. O încercare de sistematizare a compuşilor aromatici monociclici. Studii şi Cercatări Chim. Acad. R. P. R., 7 (1959), 257-95 (Appendix 1 in collaboration with S. Teleman).
2. A. T. Balaban and F. Harary. Chemical graphs IV (Aromaticity VI): Dihedral groups and monocyclic aromatic compounds. Rev. Roumaine Chim., 12 (1967), 1511-5.
3. J. H. Conway. Regular algebra and finite machines. Chapman and Hall (1971).
4. A. Kaufmann. Graphs, dynamic programming and finite games. Academic Press (1967).
5. R. C. Read. Contributions to the cell growth problem. Can. J. Math., 14 (1962), 1-20.

University of Southampton,
Southampton, England

ON A FAMILY OF PLANAR BICRITICAL GRAPHS

L. LOVÁSZ* and M. D. PLUMMER

1. Introduction

A 1-factor of a graph G is a set of independent lines in G which span V(G). Tutte [7] gives necessary and sufficient conditions for a graph to contain a 1-factor. A natural question to then ask is 'how many different 1-factors may a graph possess?'

Kotzig [3] and, independently, Beineke and Plummer [1] proved that any 2-connected graph with a 1-factor contains at least two of them. Zaks [8] generalized this result by proving that any k-connected graph $(k \geq 2)$ has at least $k(k-2)(k-4)\ldots$ 1-factors and this is in a sense, best possible, since the complete graphs K_{k+1} (k odd) contain exactly $k(k-2)(k-4)\ldots$ 1-factors. On the other hand, Lovász [4] has shown that many k-connected graphs have at least k! 1-factors. More precisely, any such k-connected graph with a 1-factor which is not bicritical has at least k! 1-factors. A graph is <u>bicritical</u> if for every pair of points u and v in G, $G - u - v$ has a 1-factor.

In view of the Lovász result above, it becomes natural to ask 'what is the lower bound on the number of 1-factors a bicritical graph must have?' In another paper [5], the authors obtain a number of structural results on bicritical graphs in general which do provide a lower bound $\frac{p}{4} + 2$ (p is the number of points) which is not thought to be best possible. In these studies, however, the authors were led, in particular, to the study of a subclass $\mathcal{H}$ of all bicritical graphs which seem interesting from several graph-theoretical points of view and for which a best possible lower bound has been obtained. We call $\mathcal{H}$ the class of Halin graphs.

---

* Work partially supported by NSF contract GP-29129

A graph H is a <u>Halin graph</u> if it can be constructed as follows: let T be any tree in which each non-endpoint has minimum degree 3. Embed T in the plane and construct a cycle through all the endpoints of T in such a way that $H = T \cup C$ remains plane (cf. Figure 1).

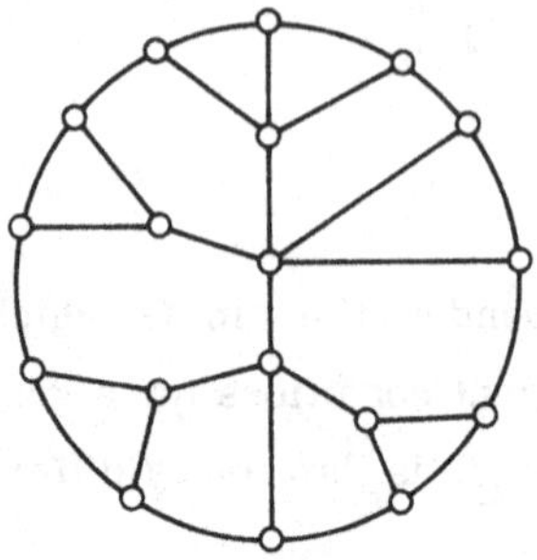

Figure 1

Why are such graphs of interest? Halin [2] uses them in his work on minimally 3-connected graphs (hence our name for $\mathcal{H}$). Moreover, the cubic Halin graphs are in 1:1 correspondence (via duality) with the triangulations of the disc. Bondy and Lovász [unpublished] proved that these graphs are almost pancyclic. They contain cycles of each length between 3 and $|V(G)|$ except, possibly, one and this one exception must be of even length.

The subclass of $\mathcal{H}$ in which the graphs are cubic have been studied by Rademacher [6] and others with the purpose of enumerating them. Rademacher calls this subclass the <u>based polyhedra</u> and Polya calls them <u>roofless polyhedra.</u> Indeed, interest in these graphs may be traced all the way back to Kirkman.

For our purposes, however, they are interesting for different reasons. Let $\phi(G)$ denote the number of different 1-factors contained in graph G. Since we seek a lower bound for $\phi(G)$ when G is bicritical, it is natural to study graphs which are bicritical and <u>minimal</u> with respect to this property; i. e. G is bicritical, but for each line x, G - x is not bicritical. We shall show below that every even Halin graph is such a minimal bicritical graph. In the final section we shall point out another possible reason for the study of class $\mathcal{H}$.

Theorem 1. (Bondy) *All Halin graphs* H *are* 1-*Hamiltonian; i.e. both* H *and* H - v *have Hamiltonian cycles for all points* v.

The preceding unpublished result of Bondy is used to prove

Theorem 2. *Every even Halin graph* H *is minimal bicritical.*

Theorem 3. *Let* G *be a graph such that there exists a partition* $V(G) = A \cup B$ *with* $|A|, |B| \geq 2$ *such that the two new graphs formed by contracting* A, B, *respectively onto one point are bicritical. Then* G *is bicritical.*

Definition. A bicritical graph G is *contraction-reducible* if there is a partition $V(G) = A \cup B$ with $|A|, |B| \geq 2$ such that if two new graphs G_A and G_B are formed by respectively contracting B to a point and A to a point, then G_A and G_B are bicritical. Otherwise, G is *contraction-irreducible.* (Cf. Figure 2.)

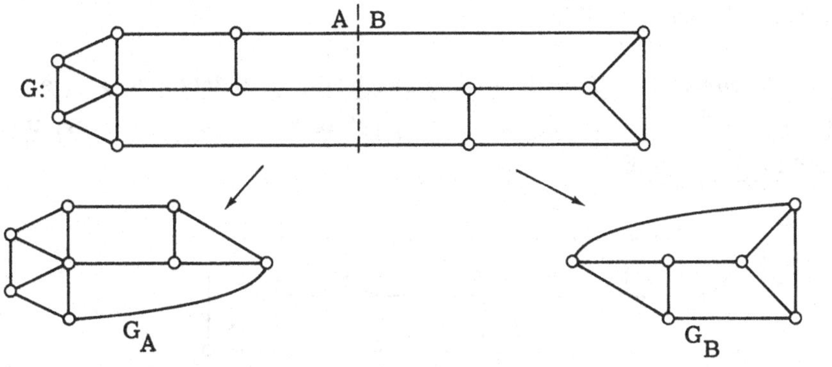

(a) G is contraction-reducible

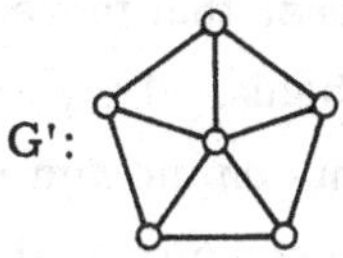

(b) G' is contraction-irreducible

Figure 2

Definition. An even Halin graph $H = T \cup C$ has the *even-cut property* if each non-endline of T separates T into two subtrees each with an even number of points.

Theorem 4. *The following are equivalent for any even Halin graph* $H = T \cup C$.

(i) *Each inner point of* T *is joined to an odd number of endpoints of* T.

(ii) H *has the even-cut property.*

(iii) H *is contraction-irreducible.*

We shall denote by $\hat{\mathcal{H}}$ the subclass of all even Halin graphs satisfying any of the three equivalent statements of the preceding theorem.

Our next objective is to determine a sharp lower bound for $\phi(H)$, when $H \in \hat{\mathcal{H}}$.

Theorem 5. *If* $H \in \hat{\mathcal{H}}$ *and* $|V(H)| = p$, *then* $\phi(H) \geq p - 1$ *and this bound is best possible.*

Theorem 6. *If* H *is a Halin graph with* p *points and* p *is even, then* $\phi(H) \geq \frac{2}{3}(p - 1)$ *unless* H *is the graph of Figure 3. Moreover, this bound is best possible.*

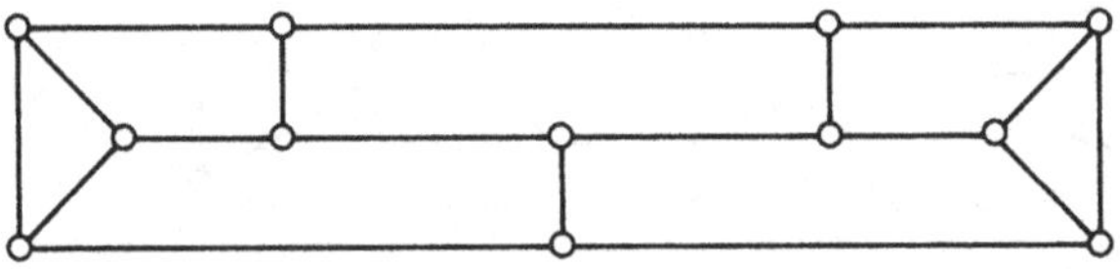

Figure 3

This result is best possible in the sense that there is an infinite family of even Halin graphs where equality holds. They are constructed as follows: let W_r be a wheel with r points on the rim each joined to a common point v where r is odd. For each point on the rim insert a triangle. (cf. Figure 4.)

Then the resulting graph W'_r has $3r + 1$ points and $2r$ 1-factors. Thus $\phi(W'_r) = \frac{2}{3}(p - 1)$.

The proofs of the above theorems will appear elsewhere.

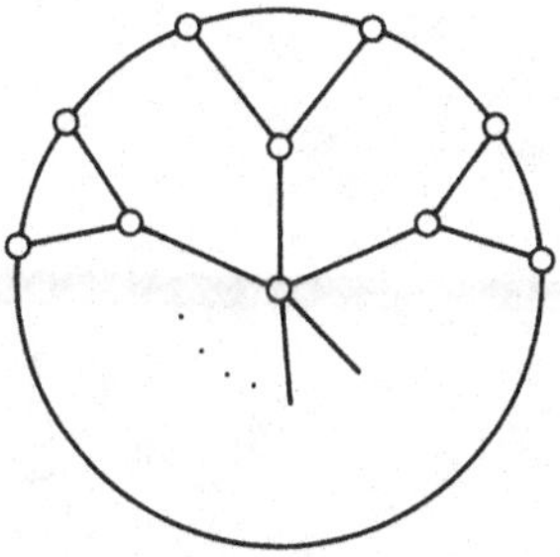

Figure 4

References

1. L. W. Beineke and M. D. Plummer. On the 1-factors of a non-separable graph. J. Comb. Theory, 2 (1967), 285-9.
2. R. Halin. Studies on minimally n-connected graphs. Combinatorial Mathematics and its Applications, D. J. A. Welsh, Ed., Academic Press, New York (1971), 129-36.
3. A. Kotzig. Ein Beitrag zur Theorie der endlichen Graphen mit linearen Faktoren I, II, III. Mat. Fyz. Časopis, 9 (1959), 73-91, 136-59, and 10 (1960), 205-15 (in Slovak, with German summary).
4. L. Lovász. On the structure of factorizable graphs. Acta Math. Acad. Sci. Hung., 23 (1972), 179-95.
5. L. Lovász and M. D. Plummer. On bicritical graphs. Submitted for publication.
6. H. Rademacher. On the number of certain types of polyhedra. Illinois J. Math., 9 (1965), 361-80.
7. W. T. Tutte. The factorization of linear graphs. J. London Math. Soc., 22 (1947), 107-11.
8. J. Zaks. On the 1-factors of n-connected graphs. Combinatorial structures and their applications, Gordon and Breach, New York (1970), 481-8.

Eötvös L. University,
Budapest, Hungary

Vanderbilt University,
Nashville, Tennessee, U. S. A.

ON THE ENUMERATION OF PARTIALLY ORDERED SETS OF INTEGERS

J. C. P. MILLER

1. Introduction

The problem originally posed was the enumeration of sets of ten integers, placed in a triangular array as in Fig. 1, with the restrictions:

(1) Each number is less than any number on a lower level, as indicated in the directed graph in Fig. 2 (arrows indicate increase).

(2) The totals in the columns are respectively 4n, 3n, 2n, n = d.

We consider whether the number of possible sets can be expressed as a polynomial in n.

a	b	c	d
e	f	g	
h	i		
j			

Fig. 1

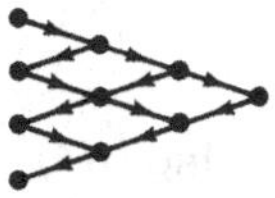

Fig. 2

1	2	3	4
3	4	5	
5	6		
7			

Fig. 3

2. Requirements

In starting from scratch we need, or hope for, several things:

(i) Counts for small n, by direct construction and counting, are needed for inspiration and checking.

(ii) Methods for speeding the counting.

(iii) Recurrence relations.

(iv) Generating functions.

(v) General expressions in n or N (the overall sum or <u>weight</u>) for the counts.

The overall sum N = 10n is here rather large for each initial counting and we will first consider a simpler problem: (a) we first cancel the second condition so that only the first holds, (b) next we replace 'less than' by 'not greater than'; this is equivalent to the subtraction of the minimum non-zero set satisfying (1), see Fig. 3. (c) We start with a triangle of six elements only (Fig. 4).

3. Initial counts

Consider then a triangle of six elements, with condition (1'), i. e. condition (1) modified to allow equality of elements. We readily construct

```
a
   b
d      c
   e
f
```

Fig. 4

and count sets of weight $N = 1(1)5$, using a dot for a zero element. The number of sets of weight N is here denoted by $f(N)$.

```
N = 0   | N = 1   | N = 2          | N = 3
        |         |                |
.       | .       | .      .       | .      .      .      .
   .    |    .    |    .      .    |    .      .      1      .
.     . | .     . | .     . .  1 . | .     . .  1 . 1  1 . .  1  1
   .    | 1  .    | 2 ..      1    | 3  .      2      1      1
.       |         |                |
```

$f(0) = f(1) = 1$, $f(2) = 2$, $f(3) = 4$. Also $f(4) = 6$, $f(5) = 9$.

4. Unit sets and shapes

We will show in §7 how to build up an array satisfying (1') from the following set of patterns of units, or unit sets.

```
Set or
shape      A         B         C         D         E         F         G
                                                                  1
           .         .         .         .         .        .    1
              .         .         .         .        .       1       1
           .     .   .     .   1     .   .     1   1    1   1    1   1     1
              .         1         1         1        1       1       1
           1         1         1         1         1        1    1
Weight     1         2         3         3         4         5         6
```

(4. 1)

We shall say that any set satisfying (1') has a shape given by the arrangement of its subset of equal largest integers, e. g. it has shape D if $f = e = c > d$ or b. The shape will be called better than D if $f = e = c = d$ or $f = e = c = b$ (or both). Thus the shape will be called X or better if the subset of largest elements includes all those having a

1 in the unit set X, and exactly X if <u>only</u> those elements are included. We see that a set of weight N is the sum of some unit set X of weight w and a set of weight N - w, having shape X or better.

5. Recurrence relations

We now obtain recurrence relations by counting sets of various shapes. Denote by x(N) the number of sets of weight N and exact shape X, and by X(N) the number of sets of shape X or better. Two sets of relations follow immediately:

$$\left.\begin{aligned} a(N) &= A(N-1) \\ b(N) &= B(N-2) \\ c(N) &= C(N-3) \\ d(N) &= D(N-3) \\ e(N) &= E(N-4) \\ f(N) &= F(N-5) \\ g(N) &= G(N-6) \end{aligned}\right\} \quad (5.1)$$

$$\left.\begin{aligned} G(N) &= g(N) \\ F(N) &= f(N) + G(N) \\ E(N) &= e(N) + F(N) \\ D(N) &= d(N) + E(N) \\ C(N) &= c(N) + E(N) \\ B(N) &= b(N) + c(N) + D(N) \\ &= b(N) + C(N) + d(N) \\ A(N) &= a(N) + B(N) \end{aligned}\right\} \quad (5.2)$$

The first set (5. 1) comes from detaching the unit set X from each set having exact shape X. The second set (5. 2) comes from noting that sets with shape X or better include all those with exact shape X', where X' is in turn each shape X or better, and then reducing the relations by using those higher in the list.

Since c(0) = d(0) = 0, it follows that C(N) = D(N) and we may combine them to give

$$\gamma(N) = C(N) + D(N) = 2C(N). \qquad (5.3)$$

We note that sets C and D are mutually exclusive in our construction; this avoids double counting that might result from the pattern relation

$$C + D = \overset{\bullet}{\underset{2}{1}}\ \overset{\bullet}{{}_{2}}\ 1 = B + E$$

Eliminating a(N), b(N), c(N), d(N), e(N), f(N), g(N) and C(N), D(N) we obtain finally the effective recurrence relations

$$\left.\begin{aligned}G(N) &= G(N-6)\\ F(N) &= F(N-5) + G(N)\\ E(N) &= E(N-4) + F(N)\\ \gamma(N) &= \gamma(n-3) + 2E(N)\\ B(N) &= B(N-2) + \gamma(N) - E(N)\\ f(N) = A(N) &= A(N-1) + B(N)\end{aligned}\right\} \quad (5.4)$$

6. Sample table from recurrence

We give a sample table, obtained from these recurrence relations.

Table 1

N	A(N)	B(N)	γ(N)	E(N)	F(N)	G(N)
0	1	1	2	1	1	1
1	1	.	.	.	.	.
2	2	1	.	.	.	.
3	4	2	2	.	.	.
4	6	2	2	1	.	.
5	9	3	2	1	1	.
6	14	5	4	1	1	1
7	19	5	2	.	.	.
8	27	8	4	1	.	.
9	37	10	6	1	.	.
10	49	12	6	2	1	.

The whole table starts with $G(0) = g(0) = 1$ and is symmetric when continued backwards, with large blocks of zeros, 17 in $A(N)$, for small negative values of N. This table was extended rapidly by hand to $N = 70$ to provide more than a full cycle ($60 =$ L. C. M. of 2, 3, 4, 5) of results.

7. Unique decomposition of sets

We now consider the unique decomposition of a set into unit sets. We organise this as a flow diagram in words. We may proceed from low to high weights, as in §4, or vice versa. We choose the latter, and start with G.

1. Detach G repeatedly until $a = 0$, η times in all; → 2.
2. Detach F repeatedly until $b = 0$, ζ times in all; → 3.
3. Detach E repeatedly until $d = 0$ or $c = 0$, ε times in all, → 4 if $d = 0$, or → 5 if $c = 0$.
4. Detach D repeatedly until $c = 0$, δ times in all; → 6.
5. Detach C repeatedly until $d = 0$, γ times in all; → 6.

6. Detach B repeatedly until $e = 0$, β times in all; → 7.

7. Detach A repeatedly until $f = 0$, α times in all; END.

Decomposition is complete, and we have uniquely the set

$$\alpha A + \beta B + \gamma C + \delta D + \varepsilon E + \zeta F + \eta G \quad \text{with } \gamma\delta = 0. \tag{7.1}$$

There is only one possible unit shape of each weight allowed in this decomposition.

8. Generating function for counts. Values of f(N)

When enumerating sets with total weight N, we tag each set of weight i with a tag t^i, and use products to combine weights. For the set (7.1) we thus get the weight

$$t^{\alpha+2\beta+3\gamma+3\delta+4\varepsilon+5\zeta+6\eta}.$$

Then the full generating function is clearly

$$\begin{aligned} G(t) &= (1+t+t^2+\dots)(1+t^2+t^4+\dots)(1+2t^3+2t^6+\dots)(1+t^4+t^5+\dots) \\ &\quad (1+t^5+t^{10}+\dots)(1+t^6+t^{12}+\dots) \\ &= \frac{1}{1-t}\cdot\frac{1}{1-t^2}\cdot\frac{1+t^3}{1-t^3}\cdot\frac{1}{1-t^4}\cdot\frac{1}{1-t^5}\cdot\frac{1}{1-t^6} \end{aligned} \tag{8.1}$$

of which the coefficient of t^N, in the expansion in powers of t, gives f(N), the number of sets of weight N.

The last factor, for example, corresponds to the unit set G, when used 0, 1, 2, ... times; the third factor corresponds to unit sets C and D, we can choose neither in one way, or 1, 2, 3, ... of them in two ways each, since they cannot occur together.

Now, the generating function for p(n, m), the number of partitions of n into exactly m parts, or alternatively into parts with largest at most m is

$$G_m(t) = \sum_n p(n, m)t^n = \prod_{r=1}^{m} (1 - t^r)^{-1} \tag{8.2}$$

whence

$$f(N) = p(N, 6) + p(N - 3, 6). \tag{8.3}$$

The function p(n, m) is tabulated in <u>Tables of partitions</u>, Royal Society Mathematical Tables, Volume 4 (1958). From the introduction to these tables, we may also obtain expansions for f(N) in powers of N, or preferably in terms of binomial coefficients. It may be seen directly that f(N) is a polynomial in N.

9. The example of Figs. 1 and 2, with restriction (1')

We need a rather more complicated example to indicate in fuller detail the development of a generating function from a larger comparibility graph. This diagram, derived from the set and graph of Figs. 1

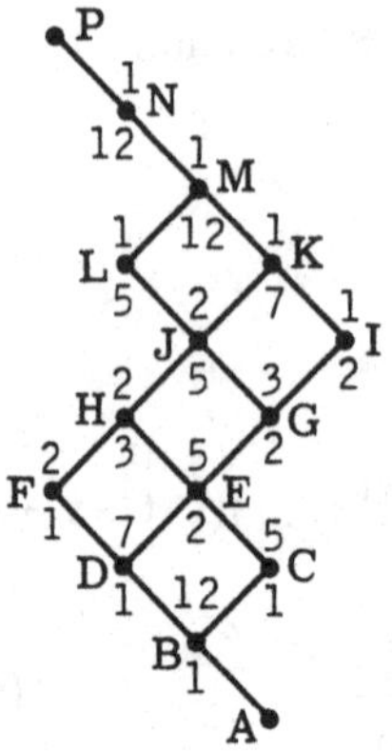

Fig. 5

and 2, and describing the partial ordering on the set of shapes is given in Fig. 5. The unit shapes are listed below.

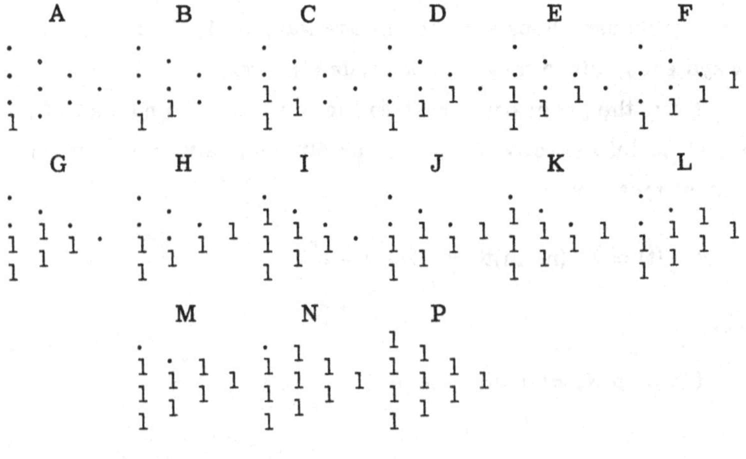

In Fig. 7 the integer above each point X gives the number of paths through the graph from the top P to that point D, and the number below counts the paths from the point X to the bottom A. The product of the two numbers at the point X gives the number of paths from P to A down the graph and going through the point X. There are 12 routes down the graph, each involving a complete set of ten patterns, one unit-pattern of each weight 10 to 1; any number of each pattern on the path 0, 1, 2, ... may occur.

10. Development of the generating function

We examine two ways of developing a generating function - the first special to this case, still relatively simple, the second very general.

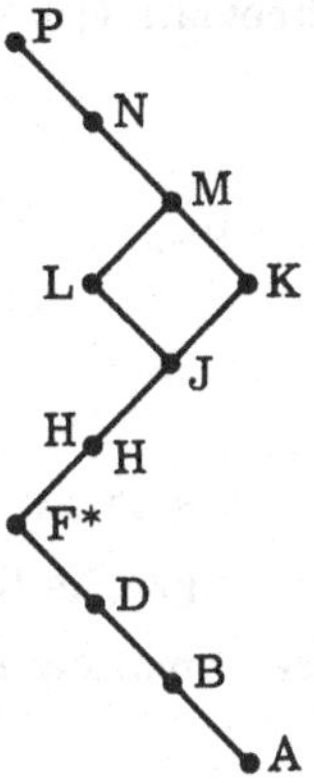

Fig. 6a

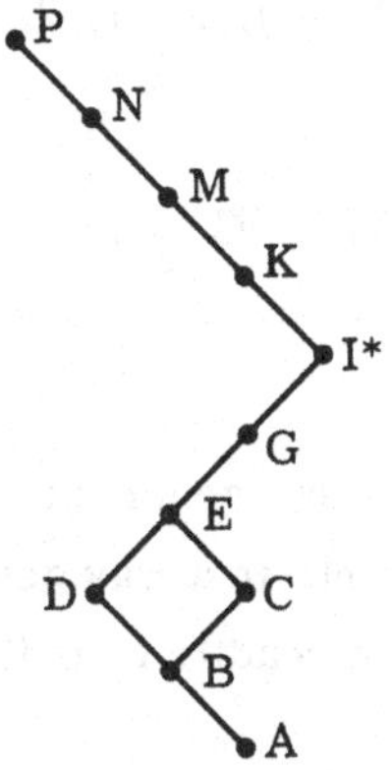

Fig. 6b

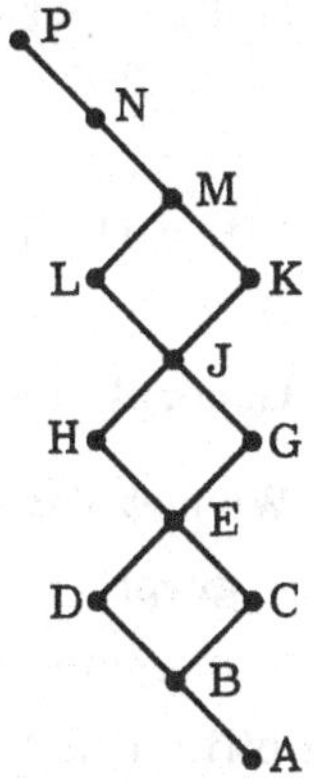

Fig. 6c

Special method. Fig. 5 may be divided into three parts that may be dealt with separately, Figs. 6a, 6b, 6c.

It is readily seen that if F is present in a set obtained from one of the two paths in Fig. 6a, the set occurs just once, but if it is absent, any such set is also given by Fig. 6c; likewise for I in Fig. 6b. Thus asterisked items must occur at least once. Again, as in §8, each separate lozenge gives, for the corresponding mutually exclusive pair, of weight w say, a factor

$$1 + 2t^{w} + 2t^{2w} + \ldots = (1 + t^{w})/(1 - t^{w}). \qquad (10.1)$$

Thus

$$G(t) = G_a(t) + G_b(t) + G_c(t)$$

where

$$G_a(t) = t^4(1 + t^7)/(1 - t)(1 - t^2)\dots(1 - t^{10}), \qquad (10.2)$$

$$G_b(t) = t^6(1 + t^3)/(1 - t)(1 - t^2)\dots(1 - t^{10}), \qquad (10.3)$$

$$G_c(t) = (1 + t^7)(1 + t^5)(1 + t^3)/(1 - t)(1 - t^2)\dots(1 - t^{10}). \qquad (10.4)$$

So

$$G(t) = G_{10}(t).\ (1+t^3+t^4+t^5+t^6+t^7+t^8+t^9+t^{10}+t^{11}+t^{12}+t^{15}) \qquad (10.5)$$

where $G_i(t) = 1/(1 - t)\dots(1 - t^i)$, $i = 1, 2, \dots$ We abbreviate (10.5) to

$$G(t) = (1..111111111..1)G_{10}(t). \qquad (10.6)$$

11. General method

We now develop the same generating function from the same compatibility graph, point by point, in a way general to all such graphs that are two-dimensional. It can readily be extended to more complicated compatibility graphs.

In general, for a planar compatibility graph (such as in Fig. 5) we consider a single quadrilateral, as in Fig. 7. We suppose that $G_U(t)$, $G_V(t)$, $G_W(t)$ are generating functions for U, V, W that count

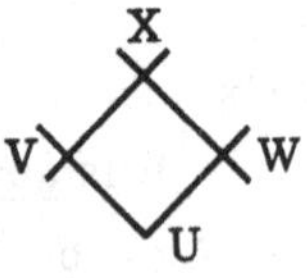

Fig. 7

every set up to those points exactly once, and that the unit set X has weight w(X). Then

$$G_V(t) + G_W(t)$$

counts every set that reaches, but does not include, X just once, except for those sets from paths that pass through U, and which do not include either of the unit sets V or W at all; these are counted twice, once via point V and once via point W. The sets that are counted twice are exactly included in the generating function $G_U(t)$. Any other set (from a planar graph with top and bottom point) that might be counted twice must miss U and yet arrive by two paths through V (or two paths through W), and this possible double count has, by hypothesis, already been rectified at V (or W). Hence

$$G_X(t) = \{G_V(t) + G_W(t) - G_U(t)\}/(1 - t^{w(X)}) \qquad (11.1)$$

We start from point A in Fig. 5 and work upwards:

$$G_A(t) = (1 + t + t^2 + \dots) = 1/(1 - t) = G_1(t) \qquad \text{(see (8.2))}.$$

$$G_B(t) = G_A(t).(1 + t^2 + t^4 + \dots) = G_1(t)/(1 - t^2) = G_2(t).$$

$$G_C(t) = G_D(t) = G_B(t).(1 + t^3 + t^6 + \dots) = G_2(t)/(1 - t^3) = G_3(t).$$

$$G_F(t) = G_D(t)/(1 - t^4) = G_4(t).$$

$$G_E(t) = \{G_C(t) + G_D(t) - G_B(t)\}/(1 - t^4).$$

$$G_H(t) = \{G_E(t) + G_F(t) - G_D(t)\}/(1 - t^5)$$
$$= G_5(t).(1\ .\ .\ 1\ 1)$$

$$G_G(t) = G_E(t)/(1 - t^5) = G_5(t).(1\ .\ .\ 1\ 1)$$

$$G_J(t) = \{G_G(t) + G_H(t) - G_E(t)\}/(1 - t^6)$$
$$= G_5(t).(1\ .\ .\ 1\ 1\ 1\ .\ .\ 1)$$

$$G_I(t) = G_G(t)/(1 - t^6) = G_6(t).(1\ .\ .\ 1)$$

$$G_K(t) = \{G_J(t) + G_I(t) - G_G(t)\}/(1 - t^7)$$
$$= G_7(t).(1\ .\ .\ 1\ 1\ 1\ 1\ .\ 1\ 1)$$

$$G_L(t) = G_J(t)/(1 - t^7) = G_7(t).(1\ .\ .\ 1\ 1\ 1\ .\ .\ 1)$$

$$G_M(t) = \{G_K(t) + G_L(t) - G_J(t)\}/(1 - t^8)$$
$$= G_8(t).(1\ .\ .\ 1\ 1\ 1\ 1\ 1\ 1\ 1\ 1\ 1\ 1\ .\ .\ 1)$$

$$G_P(t) = G_N(t)/(1 - t^{10}) = G_M(t)/\{(1 - t^9)(1 - t^{10})\}$$
$$= G_{10}(t).\,(1\,.\,.\,1\,1\,1\,1\,1\,1\,1\,1\,1\,1\,.\,.\,1)$$

agreeing with (10. 4).

The whole process may be carried out easily in terms of the numerators $g_X(t)$ only, where

$$G_X(t) = G_{W(X)}(t)\,.\,g_X(t).$$

We will not here investigate the derivation of a recurrence relation for $g_X(t)$ in individual special cases.

12. Results for Figs. 1, 2 with condition (1')

The generating function (10. 4) gives directly

$$f(N) = p(N,\ 10) + \sum_{N=3}^{12} p(N - r,\ 10) + p(N - 15,\ 10) \qquad (12.1)$$

which can be used in conjunction with R. S. Math. Tab. 4 (1958).

Values of f(N) for N = 0(1)15 are given in Table 2.

Table 2

N	f(N)	N	f(N)	N	f(N)	N	f(N)
0	1	4	7	8	41	12	172
1	1	5	11	9	60	13	235
2	2	6	18	10	87	14	320
3	4	7	27	11	122	15	430

13. A partially-ordered set with symmetries

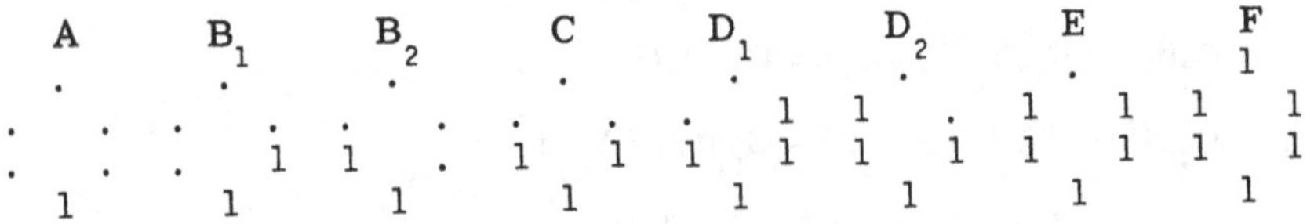

The points and order-relations are given in Figs. 8 and 9, with the compatibility graph in Fig. 10, using the unit shapes given alongside with their weights. Fig. 10 alone is sufficient for the development of the generating function

a
b c
d e
f

Fig. 8

Fig. 9

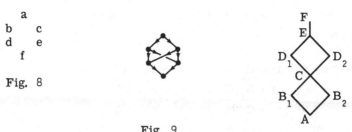

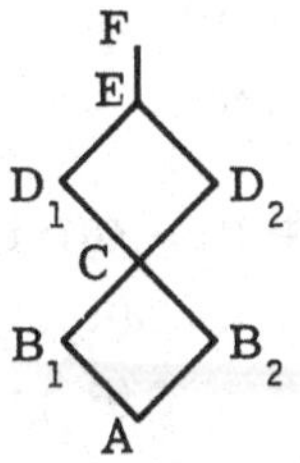

Fig. 10

$$G_F(t) = \frac{1}{1-t} \cdot \frac{1+t^2}{1-t^2} \cdot \frac{1}{1-t^3} \cdot \frac{1+t^4}{1-t^4} \cdot \frac{1}{1-t^5} \cdot \frac{1}{1-t^6} = G_6(t).\,(1+t^2)(1+t^4) \tag{13.1}$$

This is the generating function for counting sets when pairs of distinct sets that are equivalent under symmetry are each counted separately. The symmetries are R_1, which interchanges d and e, R_2 which interchanges b and c, and R_1R_2 which interchanges both pairs. It is also of interest to count patterns in which pairs equivalent under any of these symmetries are counted as one pattern only.

These symmetries form a group of permutations of the points, with four elements, and we can use Burnside's lemma to count patterns. Polya's Theorem does not apply directly; the cycle index of the group is $\frac{1}{4}(t_1^6 + 2t_1^4t_2 + t_1^2t_2^2)$, whereas the generating functions corresponding to fixed sets under the symmetries R_1 and R_2 are distinct, though each gives a term $t_1^4t_2$ in the cycle index.

For use with Burnside's lemma, we need generating functions:

$G(R_1; t)$ for sets unchanged under R_1
$G(R_2; t)$ for sets unchanged under R_2
$G(R_1R_2; t)$ for sets unchanged under R_1R_2.

Now R_1 interchanges d and e, i.e. B_1 and B_2 and, since they cannot occur at all if the set is unaltered under R_1, so the factor $(1 + t^2)/(1 - t^2)$ becomes $1 = (1 - t^2)/(1 - t^2)$.

Likewise no change under R_2 means that D_1 and D_2 are absent and $(1 + t^4)/(1 - t^4)$ is replaced by $1 = (1 - t^4)/(1 - t^4)$.

Finally under R_1R_2 both factors are altered.

Hence

$$G_F(t) = G(I; t) = G_6(t) \, . \, (1 + t^2 + t^4 + t^6) \qquad (13.1)$$

$$G(R_1; t) = G_6(t) \, . \, (1 - t^2 + t^4 - t^6) \qquad (13.2)$$

$$G(R_2; t) = G_6(t) \, . \, (1 + t^2 - t^4 - t^6) \qquad (13.3)$$

$$G(R_1R_2; t) = G_6(t) \, . \, (1 - t^2 - t^4 + t^6) \qquad (13.4)$$

Hence the total number of patterns is counted by

$$P(t) = \tfrac{1}{4}\{G(I; t) + G(R_1; t) + G(R_2; t) + G(R_1R_2; t)\}$$
$$= G_6(t) = \sum p(N, 6)t^N.$$

So the number of patterns of weight N is exactly equal to the number of partitions of N into exactly six parts.

14. A further example

We now examine an example, still with symmetry, but having a non-planar compatibility graph (Figs. 11, 12, 13). The corresponding unit sets are now listed:

a
b d
c
f
g e
h

Fig. 11

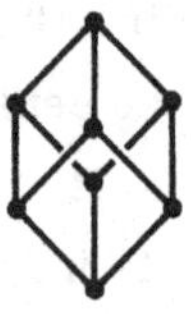

Fig. 12

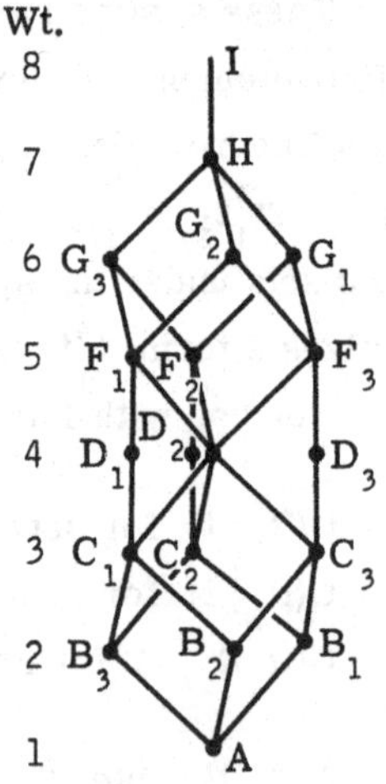

Fig. 13

	A			B			C			D			E	
				(B_1,B_2,B_3)			(C_1,C_2,C_3)			(D_1,D_2,D_3)				
	.			.			.			.			.	
.		.	.		.	.		.	1		.	.		.
	.			.			.			.			.	
	.			.			1			1			1	
.		.	.		1	1		.	1		.	1		1
	1			1			1			1			1	

	F			G			H			I	
	(F_1,F_2,F_3)			(G_1,G_2,G_3)							
	.			.			.			1	
1		.	.		1	1		1	1		1
	.			1			1			1	
	1			1			1			1	
1		1	1		1	1		1	1		1
	1			1			1			1	

The generating functions, developed as in §11 by working up the compatibility graph, are

$G_A(t) = G_1(t)$
$G_B(t) = G_2(t)$ all equal for B_1, B_2, B_3
$G_C(t) = G_3(t).(1\ .\ 1)$ all equal
$G_D(t) = G_4(t).(1\ .\ 1)$
$G_E(t) = G_4(t).(1\ .\ 2\ 2\ .\ 1)$ *see comment below
$G_F(t) = G_5(t).(1\ .\ 2\ 2\ 1\ 1\ 1)$
$G_G(t) = G_6(t).(1\ .\ 2\ 2\ 2\ 2\ 2\ 2\ 2\ .\ 1)$
$G_H(t) = G_7(t).(1\ .\ 2\ 2\ 3\ 3\ 5\ 4\ 8\ 4\ 5\ 3\ 3\ 2\ 2\ .\ 1)$ *see comment below

$G(t)=G_I(t)=G_8(t).(1\ .\ 2\ 2\ 3\ 3\ 5\ 4\ 8\ 4\ 5\ 3\ 3\ 2\ 2\ .\ 1)$

* These two functions involve the meeting of three paths. We use the familiar method of inclusion and exclusion as in §11.

The group of symmetry permutations on the points of the graph in Fig. 12 is S_3 and the cycle-index is $\frac{1}{6}(t^8 + 2t_1^2t_3^2 + 3t_1^4t_2^2)$, corresponding to the identity, two rotations, ±120°, and three reflexions. For the rotations B_1, B_2, B_3 must be treated identically, and so all are absent (since they are mutually exclusive) if no change is to occur; likewise for all C, D, F, G. Hence only A, E, H, I, can occur and the generating function for arrangements unaltered by either of these rotations is

$$G_{rot}(t) = \frac{1}{1-t} \cdot \frac{1}{1-t^4} \cdot \frac{1}{1-t^7} \cdot \frac{1}{1-t^8}$$
$$= G_8(t).(1-t^2)(1-t^3)(1-t^5)(1-t^6)$$
$$= G_8(t).(1\ .\ \bar{1}\,\bar{1}\ .\ .\ \bar{1}\,1\,2\,1\,\bar{1}\ .\ .\ \bar{1}\,\bar{1}\ .\ 1)$$

For the reflexion R_1 which leaves X_1 unmoved, and interchanges X_2 and X_3, we must have, for $X_1 + B_1$ and no change in the set, B_2 and B_3 equally and so absent, likewise for C, D, F, G. So, with this symmetry, only A, B_1, C_1, D_1, E, F_1, G_1, H, I can occur, giving the compatibility graph in Fig. 14. Thus, working through,

$$G_{ref}(t) = G_8(t).(1\ .\ .\ .\ 1\,1\,1\ .\ .\ .\ 1\,1\,1\ .\ .\ .\ 1).$$

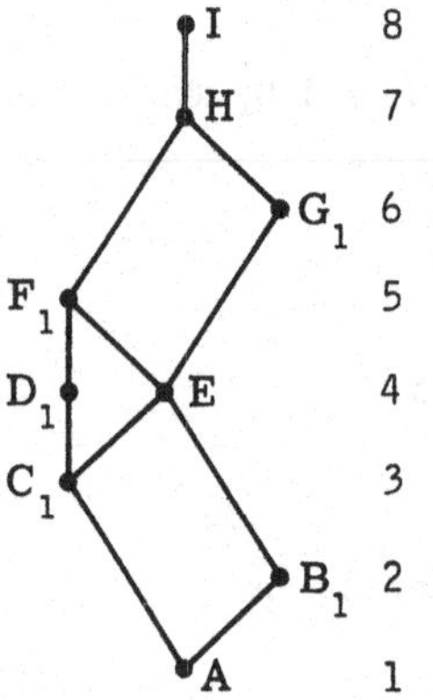

Fig. 14

The generating function for distinct patterns is then, by Burnside's lemma

$$P(t) = \tfrac{1}{6}\{G(t) + 2G_{rot}(t) + 3G_{ref}(t)\}$$
$$= G_8(t).(1\ .\ .\ .\ 1\ .\ .\ 1\,2\,1\ .\ .\ 1\ .\ .\ .\ 1)$$
$$= G_8(t).(1 + t^4 + t^7 + 2t^8 + t^9 + t^{12} + t^{16}),$$

and the coefficient of t^N is

$$P(N) = p(N,\ 8) + p(N-4,\ 8) + p(N-7,\ 8) + 2p(N-8,\ 8)$$
$$+ p(N-9,\ 9) + p(N-12,\ 8) + p(N-16,\ 8).$$

Table 3 gives the first few coefficients in the four functions.

Table 3

N	0	1	2	3	4	5	6	7	8	9	10
$G(t)$	1	1	4	7	14	23	41	63	104	152	230
$G_{ref}(t)$	1	1	2	3	6	7	11	15	22	28	38
$G_{rot}(t)$	1	1	1	1	2	2	2	3	5	5	5
$P(t)$	1	1	2	3	6	8	13	19	30	41	59

The sets so far considered have all been related to partitions of N in various ways. The next paragraph considers sets with a given final element, where results are represented in terms of binomial coefficients.

15. Sets with a fixed largest element

These are readily counted in a similar fashion. In fact, this largest element is precisely the number of unit-shapes involved in a set. If the largest element is n, and we add to the compatibility graph a loop at each vertex representing a shape, the count gives the number of paths up the graph that pass round exactly n loops.

For the set of Fig. 4, the generating function (8.1) is replaced by

$$G(t) = \frac{1}{1-t}\cdot\frac{1}{1-t}\cdot\frac{1+t}{1-t}\cdot\frac{1}{1-t}\cdot\frac{1}{1-t}\cdot\frac{1}{1-t} = \frac{1+t}{(1-t)^6}$$

giving each unit set a weight 1. Thus the number with largest element n is the coefficient $l(n)$ in the above, namely

$$l(n) = \binom{n+5}{5} + \binom{n+4}{5}$$

For the set of Fig. 1, I find

$$G(t) = (1 + 5t + 5t^2 + t^3)/(1 - t)^{10}$$

and

$$l(n) = \binom{n+9}{9} + 5\binom{n+8}{9} + 5\binom{n+7}{9} + \binom{n+6}{9}.$$

We note that $l(1)$ is the number of unit sets.

We may remark that there is a further symmetry possibly in these sets. If every element is subtracted from the sum of the first and last, we derive another solution with identical end elements, which may or may not match the original set throughout.

16. Concluding remarks

I have described in rather full detail the several examples exhibited in this paper because its purpose is a practical one. It is designed to show an effective method for deriving generating functions and counting polynomials for various kinds of arrangements of integers that can be based on partially ordered sets, with considerable emphasis on obtaining actual numerical results for small sets.

Several aspects could profit from more investigation. For example the derivation and use of recurrence relations needs development, particularly in the practical use of circulators, which give a convenient real form for counts depending on complex solutions of characteristic equations. Another area that needs developing is the organisation of simple counts just beyond the possibility of complete listing of examples.

Several of the results obtained have been expressed in terms of partition counts. For most of these simple binomial expressions are not available, and numerical values must usually be obtained from tables or by use of recurrence, so that organisation of these calculations along lines similar to those in this paper offer one of the best approaches. I am grateful to D. M. Jackson for encouraging me to look into this problem and providing a particular case to start on. Lack of space forbids the inclusion of the solution of the original problem presented in §1. This problem has been solved completely. Its solution will, I hope, appear, with an account of its origin, in a paper with D. M. Jackson. Several other similar problems come to mind involving plane and higher dimensional partitions, cyclic partitions, paths through a lattice, and so on.

7 de Freville Avenue,
Cambridge, England

THE DISTANCE BETWEEN NODES IN RECURSIVE TREES

J. W. MOON

1. Introduction

A tree is a connected graph that has no cycles (see [2] or [5] for definitions not given here). A tree T_n with n nodes labelled $1, 2, \ldots, n$ is a recursive tree if $n = 1$ or if $n \geq 2$ and T_n is obtained by joining the nth node to one node of some recursive tree T_{n-1}. There are $(n - 1)!$ recursive trees T_n and a tree with n labelled nodes is recursive if and only if the labels of the nodes in the path from the 1st node to the kth node of the tree form an increasing subsequence of $\{1, 2, \ldots, k\}$ for $k = 1, 2, \ldots, n$.

The distance between nodes i and j of a tree is the number D_{ij} of edges in the path joining i and j. Our main object here is to determine the mean and variance of D_{ij} over the set of the $(n - 1)!$ recursive trees T_n. It follows from our results, for example, that the average distance between two nodes in a random recursive tree T_n is approximately $2 \log n$ when n is large. The corresponding result with respect to the set of the n^{n-2} ordinary trees with n labelled nodes was shown to be $(\frac{1}{2}\pi n)^{\frac{1}{2}}$ in [1]; thus, in a sense, recursive trees are shorter and bushier than ordinary trees on the average.

2. A lemma

If $1 \leq i, j \leq n$ and $0 \leq d \leq n - 1$, let $P(i, j; d)$ denote the probability that $D_{ij} = d$ in a random recursive tree T_n; notice that $P(i, i; d)$ equals 1 or 0 according as $d = 0$ or $d \geq 1$. We adopt the convention that $P(i, j; d) = 0$ if $d \geq n$ or $\max(i, j) > n$.

Suppose $i < j$; let T_{j-1} denote the subtree determined by the first $j - 1$ nodes of a recursive tree T_n and suppose the jth node of T_n is joined to node x of T_{j-1}. Then the distance between nodes i and j in T_n equals one plus the distance between nodes i and x in

T_{j-1}. The node x could be any one of the $j - 1$ nodes of the tree T_{j-1} and these possibilities are all equally likely. These considerations imply the following recurrence relation for $P(i, j; d)$.

Lemma. _If_ $1 \le i < j \le n$ _and_ $1 \le d \le n - 1$, _then_

(1) $$P(i, j; d) = \frac{1}{j-1}\{P(1, i; d-1) + P(2, i; d-1) + \ldots + P(j-1, i; d-1)\}.$$

3. The expected value of D_{ij}

If $1 \le i, j \le n$, let $E(i, j)$ denote the expected value of D_{ij} over the $(n - 1)!$ recursive trees T_n. Let $h_1 = 0$ and $h_k = \frac{1}{2} + \frac{1}{3} + \ldots + \frac{1}{k}$ for $k \ge 2$.

Theorem 1. _If_ $1 \le i < j \le n$, _then_

$$E(i, j) = h_i + h_{j-1} + 1/i.$$

Proof. If we multiply the two sides of equation (1) by d and $(d - 1) + 1$ and then sum over d we obtain the relation

(2) $$E(i, j) = 1 + \frac{1}{j-1}\{E(1, i) + E(2, i) + \ldots + E(j - 1, i)\}.$$

Hence,

(3) $$E(i, j) = 1 + \frac{j-2}{j-1}\{E(i, j-1) - 1\} + \frac{E(j-1, i)}{j-1} = E(i, j-1) + (j-1)^{-1}$$

when $i \le j - 2$; therefore,

(4) $$E(i, j) = E(i, i+1) + (i+1)^{-1} + (i+2)^{-1} + \ldots + (j-1)^{-1}.$$

Suppose that $i \ge 2$. It follows from equation (2) that

(5) $$E(i, i+1) = 1 + \frac{1}{i}\{E(1, i) + E(2, i) + \ldots + E(i-1, i)\}.$$

When we apply relation (3) to the first $i - 2$ terms in the enclosed sum we obtain the equation

$$E(i, i+1) = 1 + \frac{1}{i}\{E(1, i-1) + E(2, i-1) + \ldots + E(i-2, i-1) + E(i-1, i) + (i-2)/(i-1)\}.$$

These last two relations imply that

(6) $$E(i, i+1) = 1 + \frac{i-1}{i}\{E(i-1, i) - 1\} + \frac{1}{i}\{E(i-1, i) + (i-2)/(i-1)\}$$
$$= E(i-1, i) + 3/i - 1/(i-1).$$

Hence,

(7) $$E(i, i+1) = 2(\frac{1}{2} + \frac{1}{3} + \ldots \frac{1}{i}) + \frac{1}{i}$$

since $E(1, 2) = 1$. The required formula for $E(i, j)$ now follows from equations (4) and (7).

The following corollaries are straightforward consequences of Theorem 1.

Corollary 1. _If_ $1 < j \leq n$ _then_ $E(1, j) = 1 + \frac{1}{2} + \ldots + 1/(j-1)$ $= \log j + O(1)$.

Corollary 2. _If_ $1 \leq i < j \leq n$ _then_ $E(i, j) = E(1, i) + E(1, j) + 2(1/i - 1)$.

Corollary 3. _If_ $1 \leq i \leq n$ _and_ $n > 1$ _let_ $A(i, n)$ _denote the average distance between node_ i _and all the other nodes of a random recursive tree_ T_n; _then_

$$A(i, n) = \frac{n}{n-1}\{E(i, n+1) - 1\}.$$

Corollary 4. _If_ $n \geq 2$ _let_ $B(n)$ _denote the average distance between pairs of distinct nodes in a random recursive tree_ T_n; _then_

$$B(n) = 2\frac{n+1}{n-1}h_n - 2 = 2 \log n + O(1).$$

4. The variance of D_{ij}

If $1 \leq i, j \leq n$, let $V(i, j)$ denote the variance of D_{ij} over the $(n-1)!$ recursive trees T_n. Let $H_1 = 0$ and $H_k = 1/2^2 + 1/3^2 + \ldots + 1/k^2$ for $k \geq 2$.

Theorem 2. _If_ $1 \leq i < j \leq n$, _then_

$$V(i, j) = E(i, j) - 3H_i - H_{j-1} + 2 - 4h_i/i - 2/i - 1/i^2.$$

Proof. Let $S(i, j)$ denote the expected value of $D_{ij}(D_{ij} - 1)$ over the $(n - 1)!$ recursive trees T_n. If we multiply the two sides of equation (1) by $d(d - 1)$ and $(d - 1)(d - 2) + 2(d - 1)$, sum over d, and appeal to equation (2), we obtain the relation

$$S(i, j) = 2E(i, j) - 2 + \frac{1}{j-1}\{S(1, i) + S(2, i) + \ldots + S(j - 1, i)\}. \tag{8}$$

Hence,

$$\begin{aligned} S(i, j) &= 2\{E(i, j - 1) + (j - 1)^{-1} - 1\} \\ &\quad + \frac{j-2}{j-1}\{S(i, j - 1) - 2E(i, j - 1) + 2\} + \frac{S(j - 1, i)}{j - 1} \\ &= S(i, j - 1) + 2\frac{E(i, j - 1)}{j - 1}, \end{aligned} \tag{9}$$

by (8) and (3), when $i \leq j - 2$; therefore,

$$S(i, j) = S(i, i+1) + 2E(i, i+1)/(i+1) + \ldots + 2E(i, j-1)/(j-1). \tag{10}$$

Suppose that $i \geq 2$. It follows from equation (8) that

$$S(i, i+1) = 2E(i, i+1) - 2 + \frac{1}{i}\{S(1, i) + S(2, i) + \ldots + S(i-1, i)\}. \tag{11}$$

When we apply relation (9) to the first $i - 2$ terms in the enclosed sum we obtain the equation

$$\begin{aligned} S(i, i+1) = \{2E(i, i+1)-2\} &+ \frac{1}{i}\{S(1, i-1)+S(2, i-1)+\ldots+S(i-2, i-1) \\ &+S(i-1, i)\} + \frac{2}{i(i-1)}\{E(1, i-1)+E(2, i-1)+\ldots+E(i-2, i-1)\}. \end{aligned}$$

If we now apply relations (6), (11), and (5) to the three parts of this expression for $S(i, i + 1)$ we find that

$$S(i, i+1) = S(i-1, i) + \frac{4}{i}E(i-1, i) + 2(1/i-1/(i-1)).$$

Hence,

$$S(i, i+1) = 4\{E(1, 2)/2 + E(2, 3)/3 + \ldots + E(i-1, i)/i\} + 2/i - 2 \tag{12}$$

since $S(1, 2) = 0$. Equations (10) and (12) imply that

$$(13) \qquad S(i, j) = 4\sum_{t=2}^{i} E(t-1, t)/t + 2\sum_{t=i+1}^{j-1} E(i, t)/t + 2/i - 2$$

when $1 \le i < j \le n$.

In order to simplify these sums we first observe that

$$E(i, t+1) = E(i, t) + 1/t$$

when $t \ge i + 1$, by (3). Hence,

$$2E(i, t)/t = E^2(i, t+1) - E^2(i, t) - 1/t^2$$

which implies that

$$(14) \qquad 2\sum_{t=i+1}^{j-1} E(i, t)/t = E^2(i, j) - E^2(i, i+1) - H_{j-1} + H_i.$$

Furthermore, it follows from Theorem 1 that

$$(15) \qquad E(t-1, t) = 2E(1, t) + (t-1)^{-1} - 2$$

for $t \ge 2$. Hence,

$$(16) \qquad \sum_{t=2}^{i} E(t-1, t)/t = 2\sum_{t=2}^{i} E(1, t)/t + \sum_{t=2}^{i} 1/t(t-1) - 2\sum_{t=2}^{i} 1/t$$

$$= E^2(1, i+1) - H_i - 2h_i - 1/i,$$

appealing to equation (14). Equation (15) and Corollary 1 imply that

$$(17) \qquad E^2(i, i+1) = 4E^2(1, i+1) - 8E(1, i+1) + 4E(1, i+1)/i + 4 - 4/i + 1/i^2$$

$$= 4E^2(1, i+1) - 8h_i + 4h_i/i - 4 + 1/i^2.$$

It now follows from equations (13), (14), (16), and (17) that

$$S(i, j) = E^2(i, j) - 3H_i - H_{j-1} - 4h_i/i + 2 - 2/i - 1/i^2.$$

When this expression for $S(i, j)$ is substituted in the equation

$$V(i, j) = S(i, j) + E(i, j) - E^2(i, j)$$

we obtain the required formula for $V(i, j)$.

Corollary 5. If $1 < j \leq n$ then

$$V(1,\ j) = \{1 + \tfrac{1}{2} + \ldots + 1/(j-1)\} - \{1 + 1/2^2 + \ldots + 1/(j-1)^2\}$$

$$= \log j + O(1).$$

5. Related problems

Suppose the nodes of a recursive tree T_n are susceptible to an infection and that if an uninfected node y is joined to an infected node x where $x < y$, then the infection spreads from x to y with probability p where $0 < p < 1$. If $1 \leq i \leq j \leq n$ and $0 \leq d \leq j - i$, let $q(i,\ j;\ d)$ denote the probability that the path from the 1st node to the jth node of T_n contains the ith node and that $D_{ij} = d$. It is easy to see that

$$q(i,\ j;\ d) = \frac{1}{j-1}\{q(i,\ i:d-1) + q(i,\ i+1:d-1) + \ldots + q(i,\ j-1:d-1)\}$$

when $d \geq 1$. If the ith node of T_n is infected and $Q(i,\ j)$ denotes the probability that the jth node receives the infection from the ith node, either directly or indirectly, then it follows that

$$(18)\qquad Q(i,\ j) = \sum_{d=1}^{j-i} p^d q(i,\ j:d) = \frac{p}{j-1}\{Q(i,\ i) + Q(i,\ i+1) + \ldots + Q(i,\ j-1)\}$$

when $1 \leq i < j \leq n$. It is not difficult to show that this implies that

$$(19)\qquad Q(i,\ j) = p\,\frac{\Gamma(i)}{\Gamma(j)} \cdot \frac{\Gamma(p+j-1)}{\Gamma(p+i)}$$

when $1 \leq i < j \leq n$. (We let $Q(i,\ i) = 1$ by definition.) Thus if $R(i,\ n)$ denotes the expected total number of nodes of T_n that receive the infection from the ith node, including the ith node itself, then

$$(20)\qquad R(i,\ n) = \sum_{j=i}^{n} Q(i,\ j) = \frac{n}{p}\,Q(i,\ n+1) = \frac{\Gamma(i)}{\Gamma(p+i)} \cdot \frac{\Gamma(p+n)}{\Gamma(n)}$$

when $1 \leq i \leq n$, by (18) and (19). Notice, in particular, that

$$(21)\qquad R(1,\ n) = \frac{1}{\Gamma(p+1)} \cdot \frac{\Gamma(p+n)}{\Gamma(n)} \sim \frac{1}{\Gamma(p+1)}\, n^p$$

as $n \to \infty$. If the 1st node of one of the n^{n-2} ordinary trees with n labelled nodes is infected and the infection spreads according to the above assumption, then it was shown in [3] and [4] that this will eventually result in a total of approximately $(1 - p)^{-2}$ infected nodes on the average when n is large.

Suppose a recursive tree T_n represents a sales-distribution network in which the 1st node represents the producer. Suppose each other person in the network has to pay a unit franchise fee to his immediate supplier; further suppose that each such person also has to pass on to his immediate supplier the fraction p of all franchise fees (or portions thereof) that come to him from people he supplies directly or indirectly; the producer, of course, keeps all amounts that come to him. If $1 \leq i \leq n$ let $N(i, n)$ denote the expected net return to the ith person with respect to the franchise fees paid by the $n - 1$ people other than the producer. A slight modification of the argument in the preceding paragraph can be used to show that

$$N(1, n) = \frac{1}{p\Gamma(p+1)} \cdot \frac{\Gamma(p+n)}{\Gamma(n)} - \frac{1}{p}$$

and

$$N(i, n) = \frac{(1-p)}{p} \cdot \frac{\Gamma(i)}{\Gamma(p+i)} \cdot \frac{\Gamma(p+n)}{\Gamma(n)} \cdot \frac{1}{p}$$

when $1 < i \leq n$.

Notice that $N(n, n) = -1$ for all values of p. If $p = 1$ then $N(1, n) = n - 1$ and $N(i, n) = -1$ for $1 < i \leq n$; if $p \to 0$ then

$$N(i, n) \to \frac{1}{i} + \frac{1}{i+1} + \ldots + \frac{1}{n-1}$$

for $1 \leq i < n$. Furthermore, if $N(i, n) = 0$ then $i \sim (1 - p)^{1/p} n$ when $0 < p < 1$, by Stirling's formula.

We remark in closing that if $1 \leq d \leq n - 1$ and $f(d, n)$ denotes the expected number of nodes j in a recursive tree T_n such that $D_{1j} = d$, then it is not difficult to see that

$$f(d, n) = \sum \frac{1}{a_1 a_2 \ldots a_d}$$

where the sum is over all sets of integers $\{a_1, a_2, \ldots, a_d\}$ such that $0 < a_1 < \ldots < a_d < n$. It follows from this that if $d = o(\log n)$ then

$$f(d, n) \sim \frac{(\log n)^d}{d!}.$$

6. Acknowledgements

I am indebted to Professor A. Meir for some helpful observations. The preparation of this paper was assisted by a grant from the National Research Council of Canada.

References

1. A. Meir and J. W. Moon. The distance between points in random trees. J. Comb. Th., 8 (1970), 99-103.
2. J. W. Moon. Counting labelled trees. Canadian Mathematical Congress, Montreal (1970).
3. J. W. Moon. The spread of blight in a random tree. Jaarverslag Die Suid-Afrikaanse Wiskundevereniging, 13 (1970), 83-8.
4. J. W. Moon. A problem on random trees. J. Comb. Th., 10 (1971), 201-5.
5. R. J. Wilson. Introduction to graph theory. Oliver and Boyd, Edinburgh (1972).

University of Alberta,
Edmonton, Canada

PARTITION RELATIONS

ROY NELSON

The term partition relation was coined by P. Erdős and R. Rado in 1956 [1]. Partition relations for cardinal numbers received a very thorough analysis culminating, in 1965, in the definitive paper by P. Erdős, A. Hajnal and R. Rado [2].

When all the sets involved are ordered, the situation becomes more complex and has been less thoroughly analysed.

The existence of a well-ordering makes it possible to define a modified partition relation, by selecting from $[T]^r$ a subset $[\![T]\!]^r$ consisting, roughly speaking, of sets of r 'consecutive' elements of T.

Suppose that T is of well-order type τ under the relation $\lessdot$, which we may indicate by writing

$$T = \{t_0, \ldots, \hat{t}_\tau\}_{\lessdot} .$$

The notation $\hat{t}_\tau$ means that t_τ is excluded. (We cannot write $\{t_0, \ldots, t_{\tau-1}\}$ in case τ should be of the 2nd kind.)

Then, by $[\![T]\!]^r$ we shall denote a set consisting of the elements (themselves sets) $\{t_i, \ldots, t_{i+r-1}\}$ for each non-negative i such that $i + r - 1 < \tau$, together with, for each ν of the 2nd kind $< \tau$ and each $j < r - 1$ such that $\nu + r - j - 2 < \tau$, all the elements $\{t_{\varepsilon_i}, \ldots, t_{\varepsilon_i+j}, t_\nu, \ldots, t_{\nu+r-j-2}\}$ for some subsequence (ε_i) cofinal with $(0, \ldots, \hat{\nu})$, (i.e. $\forall\ \delta < \nu\ \exists\ i,\ \varepsilon_i > \delta$).

If we call this the 'weak' definition of $[\![T]\!]^r$, then an alternative, 'strong' definition, would insist that the subsequence (ε_i) be $(\rho, \ldots, \hat{\nu})$ for some $\rho < \nu$. Both forms of the definition can be justified as ways of 'bridging the gaps' in a transfinite sequence. Which is to be preferred depends upon the context in which $[\![T]\!]^r$ is used.

Evidently, $[\![T]\!]^r$ is never unique for infinite T, but we shall always be concerned with the existence or non-existence of some $[\![T]\!]^r$

satisfying a particular requirement - as in the following definition.

For an order type α, ordinal numbers β, γ and non-negative integer r, the modified partition relation

$$\alpha \mapsto (\beta, \gamma)^r$$

asserts for every ordered set S of type α that if $[S]^r = X \cup Y$ then there must exist $T \subseteq S$ with

<u>either</u> $[\![T]\!]^r \subseteq X$ and $\text{tp } T = \beta$
<u>or</u> $[\![T]\!]^r \subseteq Y$ and $\text{tp } T = \gamma$.

When establishing positive relations, it is preferable to do so for the 'strong' form of $[\![T]\!]^r$, but for negative relations: $\alpha \not\mapsto (\beta, \gamma)^r$, the 'weak' form of $[\![T]\!]^r$ leads to a stronger result.

However, for either form of $[\![T]\!]^r$, we have the

Lemma. <u>For any order type</u> α, <u>ordinal numbers</u> β, γ <u>and non-negative integer</u> r, <u>if</u>

$$\alpha \not\mapsto (\beta, \gamma)^r$$

<u>then also</u>

$$\alpha \not\rightarrow (\beta, \gamma)^r.$$

Proof. Consider an ordered set S with $\text{tp } S = \alpha$ and suppose there exists a partition $[S]^r = X \cup Y$ such that $[\![T]\!]^r \subseteq X \Rightarrow \text{tp } T < \beta$ and $[\![T]\!]^r \subseteq Y \Rightarrow \text{tp } T < \gamma$. Then, a fortiori, for this same partition we have

$$[T]^r \subseteq X \Rightarrow \text{tp } T < \beta \text{ and}$$
$$[T]^r \subseteq Y \Rightarrow \text{tp } T < \gamma.$$

Hence, by definition, $\alpha \not\rightarrow (\beta, \gamma)^r$.

Our main result will follow as a corollary of the following.

Theorem 1. <u>For every well-order type</u> σ <u>of cardinality</u> $\aleph_\nu$

$\sigma \mapsto (4, \omega_\nu + 1)^3$.

Proof. Let S be any set of well-order type σ under the relation $<\cdot$, and in 1-1 correspondence with the set of ordinals less than ω_ν under the mapping μ. Define a disjoint partition $[S]^3 = X \cup Y$, $X \cap Y = \emptyset$, by assigning $\{x, y, z\}_{<\cdot}$ to X whenever $x, y, z \in S$ and $\mu(x) < \mu(y) > \mu(z)$.

Then certainly we cannot have $[\![\{w, x, y, z\}_{<\cdot}]\!]^3 \subseteq X$, since $\{w, x, y\}_{<\cdot} \in X$ only if $\mu(x) > \mu(y)$, whence $\{x, y, z\}_{<\cdot} \notin X$.

Consider a transfinite sequence $T = \{t_0, \ldots, t_{\omega_\nu}\}_{<\cdot} \subseteq S$, so that $\operatorname{tp} T = \omega_\nu + 1$. (If no such sequence exists, i.e. if $\operatorname{tp} S < \omega_\nu + 1$, then there is nothing to prove.)

We cannot have $\mu(t_m) > \mu(t_{m+1})\ \forall\, m < \omega_\nu$, and so $\exists\, m$ such that $\mu(t_m) < \mu(t_{m+1})$.

If now $\mu(t_\lambda) < \mu(t_\kappa)\,\forall\, \lambda, \kappa$ such that $m \leq \lambda < \kappa < \omega_\nu$, then there must exist $\eta < \omega_\nu$ such that $\mu(t_\lambda) > \mu(t_{\omega_\nu})$ whenever $\eta < \lambda < \omega_\nu$. For otherwise we should have $\mu(t_\lambda) < (t_{\omega_\nu})\,\forall\,\lambda : \lambda \geq m$ and $\lambda < \omega_\nu$. But the correspondence

$$\mu : \{t_\lambda \mid m \leq \lambda < \omega_\nu\} \longleftrightarrow \{\mu(t_\lambda) \mid m \leq \lambda < \omega_\nu\}$$

is biunique, and hence

$$|\{\mu(t_\lambda) \mid m \leq \lambda < \omega_\nu\}| = |\{t_\lambda \mid m \leq \lambda < \omega_\nu\}| = \aleph_\nu .$$

This is contradictory, since we also have

$$|\{\mu(t_\lambda) \mid m \leq \lambda < \omega_\nu\}| \leq |\{\mu \mid \mu < \mu(t_{\omega_\nu})\}| < \aleph_\nu.$$

Whence $\{t_\lambda, t_{\lambda+1}, t_{\omega_\nu}\} \in X\ \forall\,\lambda : \lambda \geq m, \eta$ and $\lambda < \omega_\nu$, and so $[\![T]\!]^3 \nsubseteq Y$.

Otherwise, $\exists\,\kappa > m$ such that $\mu(t_\lambda) > \mu(t_\kappa)$ for some $\lambda : m \leq \lambda < \kappa$. Consider the least such κ. If it is of the 2nd kind, then $t_{n+1} \neq t_\kappa\,\forall\, n < \kappa$ and we must have $\mu(t_n) < \mu(t_{n+1})\ \forall\ n : \lambda \leq n < \kappa$ by definition of κ as being minimal. Further, we have $\mu(t_n) > \mu(t_\kappa)\ \forall\ n : \lambda \leq n < \kappa$, by the same token, for otherwise

$\mu(t_\lambda) > \mu(t_\kappa) > \mu(t_n)$. Whence we have $\{t_n, t_{n+1}, t_\kappa\}_< \in X \ \forall$ such n and so $[\![T]\!]^3 \subseteq Y$.

If κ is of the 1st kind, then we must have $\mu(t_\theta) < \mu(t_{\kappa-1}) \ \forall \theta : m \leq \theta < \kappa - 1$ since κ is minimal.

Note that $m \neq \kappa - 1$, since $\mu(t_m) < \mu(t_{m+1})$ whereas $\mu(t_{\kappa-1}) > \mu(t_\lambda)$ by definition of κ and hence $\mu(t_{\kappa-1}) > \mu(t_\kappa)$.

Whence $\{t_\theta, t_{\kappa-1}, t_\kappa\}_< \in X \ \forall \theta : m \leq \theta < \kappa - 1$, and so certainly $[\![T]\!]^3 \subsetneq Y$ whether $\kappa - 1$ is of 1st or 2nd kind.

The counter-example thus established proves the theorem.

This result has an immediate generalization to the

Corollary. <u>For each integer</u> $r \geq 3$ <u>and for every well-order type</u> σ <u>of cardinality</u> $\aleph_\nu$

$$\sigma \not\mapsto (r + 1, \ \omega_\nu + 1)^r$$

Proof. Consider any ordered set S of type σ and let the mapping μ and a disjoint partition $[S]^3 = X \cup Y$ be defined as in the proof of theorem 1. We may then define a disjoint partition

$$[S]^r = U \cup V, \ U \cap V = \emptyset$$

by assigning $\{s_1, \ldots, s_r\}_<$ to U precisely when $\{s_{r-2}, s_{r-1}, s_r\}_< \in X$.

Evidently we cannot find $[\![\{s_0, \ldots, s_r\}_<]\!]^r \subseteq U$, since that would imply $[\![\{s_{r-3}, \ldots, s_r\}_<]\!]^3 \subseteq X$, which we know to be impossible.

Equally, we cannot find $T' = \{t_{3-r}, \ldots, t_{\omega_\nu}\}_<$ with $[\![T']\!]^r \subseteq V$ (and, of course, tp $T' = \omega_\nu + 1$), for then, with $T = \{t_0, \ldots, t_{\omega_\nu}\}_<$, we should have $[\![T]\!]^3 \subseteq Y$ and tp $T = \omega_\nu + 1$, which we know also not to occur for the particular partition defined.

This proves the corollary.

By virtue of the lemma, we may also deduce that

$$\sigma \not\rightarrow (r + 1, \ \omega_\nu + 1)^r$$

whenever $r \geq 3$ and σ is any ordinal $< \omega_{\nu+1}$.

The situation is different when $r = 2$, for then we have

Theorem 2. <u>For every ordinal number</u> σ, $\sigma \not\to (3, \omega_\nu + 1)^2$ <u>if and only if</u> $\sigma > \omega_\nu 2$.

Proof. Consider a set S of well-order type σ under the relation $<\cdot$ and suppose there exists a partition $[S]^2 = X \cup Y$ such that $\{x, y, z\}_{<\cdot} \subseteq S$ implies $[\![\{x, y, z\}_{<\cdot}]\!]^2 \not\subseteq X$ and $[\![T]\!]^2 \subseteq Y$ implies $tp\, T < \omega_\nu + 1$.

We must then have $S = L \cup R$, where

$$L = \{y \in S \mid y \cdot\!> x \in S \Rightarrow \{x, y\} \in Y\}$$

and

$$R = \{y \in S \mid y <\cdot\, z \in S \Rightarrow \{y, z\} \in Y\}.$$

If $\{y_1, y_2\}_{<\cdot} \subseteq L$, then $\{y_1, y_2\} \in Y$, by definition of $y_2 \in L$. Hence $[L]^2 \subseteq Y$. Similarly, $[R]^2 \subseteq Y$. So, by assumption, it follows that $tp\, L \le \omega_\nu$ and $tp\, R \le \omega_\nu$. Put $L' = L - (L \cap R)$, so that also $tp\, L' \le \omega_\nu$. Then S is what G. H. Toulmin calls a <u>shuffle</u> of L' and R, and so tp S is at most the natural sum of tp L' and tp R. (See [3] Theorem 1.35.)

Whence $tp\, S \le \omega_\nu + \omega_\nu = \omega_\nu 2$. We have thus established the sufficiency of the condition $\sigma > \omega_\nu 2$.

To complete the proof, it will be enough to show that

$$\omega_\nu 2 \not\to (3, \omega_\nu + 1)^2$$

This we may readily do by taking

$$S = \{0, \ldots, \widehat{\omega_\nu 2}\}_<, \quad L = \{0, \ldots, \hat{\omega}_\nu\}_<$$

and $R = \{\omega_\nu, \ldots, \widehat{\omega_\nu 2}\}_<$, then defining a disjoint partition $[S]^2 = X \cup Y$, $X \cap Y = \emptyset$ by putting $Y = [L]^2 \cup [R]^2$.

Then $X = \{\{l, r\} \mid l \in L, r \in R\}$ and so clearly we cannot have $[\![\{x, y, z\}_<]\!]^2 \subseteq X$.

Also, since $L \cap R = \emptyset$, if $[\![T]\!]^2 \subseteq Y$ we must have either $T \subseteq L$ or $T \subseteq R$. But $tp\, L = tp\, R = \omega_\nu$, and hence $tp\, T \le \omega_\nu$.

The condition $\sigma > \omega_\nu 2$ is therefore shown to be also necessary and the proof is complete.

Finally, for indecomposable ordinals we can prove rather more.

Theorem 3. If σ is an indecomposable ordinal, then

$$\sigma \mapsto (3, \sigma)^2.$$

Proof. Let S be a set of well-order type σ under the relation $\lessdot$ and consider any partition $[S]^2 = X \cup Y$.

Suppose there does not exist $\{x, y, z\}_{\lessdot} \subset S$ such that $[\![\{x, y, z\}]\!]^2 \subseteq X$. Then it must follow for every element $y \in S$ that either $y \in L$ or $y \in R$ (or both) where

$$L = \{y \in S \,|\, y \gtrdot x \in S \Rightarrow \{x, y\} \in Y\}$$

and

$$R = \{y \in S \,|\, y \lessdot z \in S \Rightarrow \{y, z\} \in Y\}$$

If $\{y_1, y_2\}_{\lessdot} \subseteq L$, then $\{y_1, y_2\} \in Y$ by definition of $y_2 \in L$. So $[L]^2 \subseteq Y$. Similarly, if $\{y_1, y_2\}_{\lessdot} \subseteq R$, then $\{y_1, y_2\} \in X$ by definition of $y_1 \in L$. So $[R]^2 \subseteq Y$.

Since $S = L \cup R$ and σ is indecomposable, we must have either tp $L \geq \sigma$ or tp $R \geq \sigma$ (or both).

In either case, we can find a set T with $[T]^2 \subseteq Y$ and tp $T = \sigma$, so the theorem is certainly true, a fortiori.

References

1. P. Erdős and R. Rado. A partition calculus in set theory. Bull. American Math. Soc., 62 (1956), 427-89.
2. P. Erdős, A. Hajnal and R. Rado. Partition relations for cardinal numbers. Acta Mathematica, XVI (1965), 93-196.
3. G. H. Toulmin. Shuffling ordinals and transfinite dimension. Proc. London Math. Soc., (3) 4 (1954), 177-95.

The Open University,
Milton Keynes, England

ON A PROBLEM OF DAYKIN CONCERNING INTERSECTING FAMILIES OF SETS

J. SCHONHEIM

Daykin asked the following question:

Let M be a set with $|M| = n$, $A_i \subset M$, $1 \le i \le s$. How large can s be, if $A_i \cap A_j \ne \emptyset$ and $A_i \cup A_j \ne M$ for each i and j?

Brace and Daykin conjectured that $s \le 2^{n-2}$. I will prove this conjecture by using a corollary of the following theorem (cf. Erdős, Herzog, Schönheim; Israel Journal 1970, 408-12):

Theorem 1. Let M be a set and let $A \subset M$. Denote the set $M - A$ by $\bar{A}$. If $F = \{A_1, A_2, \ldots, A_s\}$ is a family of distinct subsets of M with the property

(i) $X \subset A_i \Rightarrow X \in F$,

then there exists a permutation σ of $\{1, 2, \ldots, s\}$ such that

(ii) $A_i \subset \bar{A}_{\sigma(i)}$.

Corollary. If F is a family of distinct sets having the property (i) and B is a subfamily of F such that $X \in B$, $Y \in B \Rightarrow X \cap Y \ne \emptyset$, then

$|B| \le \frac{1}{2}|F|$.

This is true since by (ii) B cannot contain both A_i and $A_{\sigma(i)}$.

Theorem 2. Let $A \cup B$ be a partition of the set of subsets of a set M of size n, such that $X \in A$, $Y \in A \Rightarrow X \cap Y \ne \emptyset$ and $|A| = 2^{n-1} = |B|$. If B' is a subfamily of B such that $X \in B'$, $Y \in B' \Rightarrow X \cap Y \ne \emptyset$ then $|B'| \le 2^{n-2}$.

Proof. Observe that if G is a member of B then $\overline{G}$ is a member of A and therefore $H \subset G \Rightarrow H \in B$. So, B has the property (i) and by the corollary $|B'| \leq \frac{1}{2}|B| = 2^{n-2}$.

This theorem clearly implies the following theorem corresponding to Daykin's problem:

Theorem 3. If M is a set of size n and A_i $(1 \leq i \leq s)$ are distinct subsets of M satisfying the conditions

(iii) $A_i \cap A_j \neq \emptyset$ and $A_i \cup A_j \neq M$ for each i and j,

then $s \leq 2^{n-2}$.

Proof. First, note that $A_i \cup A_j \neq M$ is the same as $\overline{A}_i \cap \overline{A}_j \neq \emptyset$. But by a theorem of Erdős-Ko-Rado, (Quart. J. of Math. Oxford II, 12 (1961), 313-20), if $s < 2^{n-1}$ it is always possible to complete the collection $A_1, A_2, \ldots, A_s$ to $A_1, A_2, \ldots, A_s, A_{s+1}, \ldots, A_{2^{n-1}}$ without contradicting the first requirement of (iii). The result then follows from theorem 2.

Professor Erdős informed me that Lovasz has also a proof, using a theorem of Marica and myself (Canad. Bull. 12 (1969), 635-8).

On the connection between the cited Erdős-Herzog-Schönheim theorem and this theorem see the remark in Herzog-Schönheim, Discrete Mathematics, 1 (1972), 329-32.

The solution of Daykin's problem would also follow from the following conjecture. (A weaker version has been stated independently by Hilton, Proceedings of the British Combinatorial Conference, Oxford 1972, p. 35.)

Let $A_1, A_2, \ldots, A_s$ be distinct subsets of E, $|E| = n$. Denote by t the number of subsets B of E such that either $B \subseteq A_i$ for some i or $B \supset A_j$ for some j, $1 \leq i, j \leq s$, then

$$t \geq 3 \min(s, 2^{n-2}).$$

Remark. P. D. Seymour has now proved the above conjecture.

UNSTABLE TREES

JOHN SHEEHAN

Let A(G) be the automorphism group of a graph G. Let $e \in E(G)$ then $G - e$ is the spanning graph obtained from G by deleting e. Let $AG(e) \equiv A(G - e) \backslash A(G)$. G is stable ([2], [3]) if there exists $e \in E(G)$ such that $AG(e) = \emptyset$. In this talk we make some rather discursive remarks on possible infinite extensions of the following theorem.

Theorem [1]. Let T be a finite tree then T is unstable if and only if T is an arc of length greater than one or is one of the graphs in figure 1.

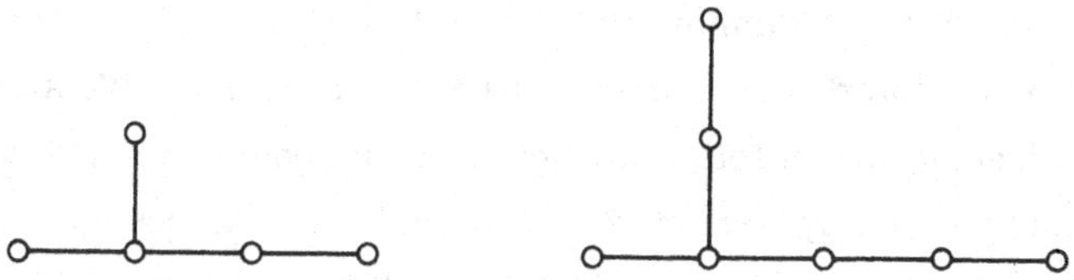

Figure 1. Unstable finite trees

Remarks. (1) Examples (Fig. 2) are easy to find of infinite unstable trees. In figure 2 there are just two vertices (indicated by broken lines) of infinite degree.

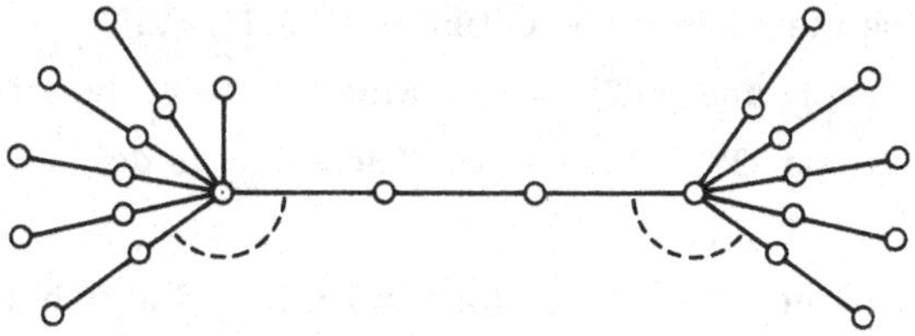

Figure 2. An infinite unstable tree

(2) If we preclude end-vertices then it is still fairly easy to give examples (Fig. 3) of infinite unstable trees. In figure 3 vertices at the $2k^{th}$ level have degree 3 and vertices at the $(2k + 1)^{th}$ level have infinite degree, $k \geq 1$.

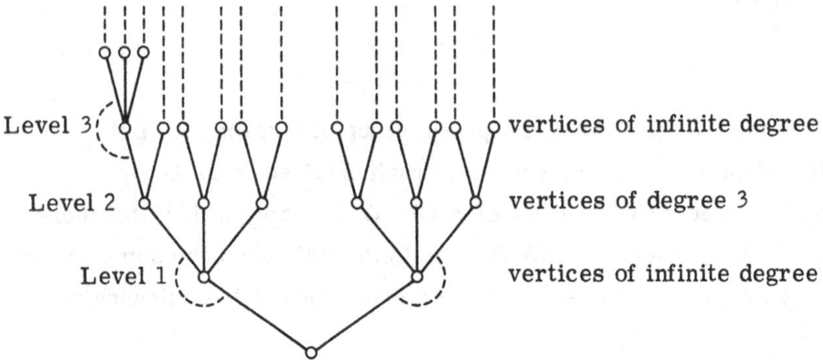

Figure 3. An infinite unstable tree without end-vertices

(3) If we restrict attention to locally finite infinite trees then we have one and only one example type of such an unstable tree. Although the construction of this example is really quite simple, it has the not unusual combinatorial defect of being difficult to describe. We attempt an informal description as follows. Let $\mathcal{T}$ be a rooted tree with root $\rho(\mathcal{T})$. Then any automorphism of $\mathcal{T}$ fixes $\rho(\mathcal{T})$. $\mathcal{T}$ is <u>root-unstable</u> if $A\mathcal{T}(e) \neq \emptyset$ for all edges e not incident with $\rho(\mathcal{T})$, where again $\mathcal{T} - e$ is regarded as a rooted tree with root $\rho(\mathcal{T})$. Let $\mathcal{T}$ and $\mathcal{T}'$ be rooted trees then we write $\mathcal{T}' \subseteq \mathcal{T}$ if $\rho(\mathcal{T})=\rho(\mathcal{T}')$ and $\mathcal{T}'$ is a $\rho(\mathcal{T})$-branch of $\mathcal{T}$. It is not difficult to prove that if $\mathcal{T}'$ is any finite rooted tree then there exists a finite root-unstable tree $\mathcal{T}$ with $\mathcal{T}' \subseteq \mathcal{T}$.

Construction. Let T_0 be any finite root-unstable tree. Write $x_0 = \rho(T_0)$. Let T_i' be the rooted tree defined by $V(T_i')=V(T_{i-1}) \cup \{x_i\}$, $E(T_i') = E(T_{i-1}) \cup \{x_{i-1}x_i\}$ and $\rho(T_i') = x_i$. Finally let T_i be a finite root-unstable tree such that $T_i' \subseteq T_i$, $i \geq 1$. Then T_ω is defined to be the union of $T_0, T_1, T_2, T_3, \ldots$

Clearly, by construction, T_ω is locally finite. Notice however that T_ω has an infinite number of end-vertices. We also have:

Result. T_ω is unstable.

Proof. Let $e \in E(T_\omega)$ then, by construction, there exists $i \geq 0$ such that $e \in E(T_i)$ and e is not incident with $\rho(T_i)$. Since T_i is root-unstable, $AT_i(e) \neq \emptyset$. Let $\sigma \in AT_i(e)$ then $\sigma(\rho(T_i)) = \rho(T_i)$. Write $\hat{\sigma}(v) = \sigma(v)$ if $v \in V(T_i)$ and $\hat{\sigma}(v) = v$ if $v \in V(T_\omega)\backslash V(T_i)$, then $\hat{\sigma} \in AT_\omega(e)$. Hence $AT_\omega(e) \neq \emptyset$ and T_ω is unstable.

(4) If we insist on the locally finite infinite tree having no end-vertices then we do not have an example of such an unstable tree and we conjecture, rather wildly, as follows.

Conjecture. Let T be a locally finite infinite tree without end-vertices then T is stable.

Incidentally notice that the disjoint union of two infinite binary trees (Fig. 4) is locally finite, infinite, without end-vertices and unstable. In figure 4 all vertices have degree 3 with the exception of the two vertices of degree 2.

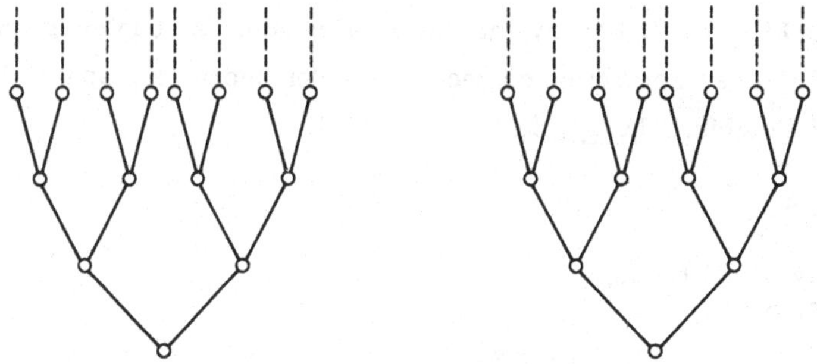

Figure 4. A locally finite unstable forest without end-vertices

(5) Finally observe that G is free ([2], [3]) if its only fixing subgraph is G itself and that if G is unstable then G is free. This is my motivation for this investigation.

References

1. N. Robertson and J. Zimmer. Automorphisms of subgraphs obtained by deleting a pendant vertex. Journal of Combin. Theory,

12 (1972), 169-73.

2. J. Sheehan. Fixing subgraphs. Journal of Combin. Theory, 12 (1972), 226-44.

3. J. Sheehan. Fixing subgraphs and Ulam's conjecture. Journal of Combin. Theory, 14 (1973), 125-36.

Appendix

In [4] a counter example consisting of a tree T and a forest 2T is given to the reconstruction conjecture for denumerable graphs. T is the point symmetric tree with denumerable vertex degree. Using the obvious vertex analogue to the concept of unstable graphs it is perhaps of interest to note that both T and 2T are vertex unstable and edge unstable.

Finally, notice that the vertex analogue of the above conjecture is false.

Reference

4. J. Fisher, F. Harary, and R. L. Graham. A counter example to the reconstruction conjecture for denumerable graphs. Journal of Combin. Theory, 12 (1972), 203-4.

University of Aberdeen,
Aberdeen, Scotland

DISTANCE-TRANSITIVE GRAPHS

D. H. SMITH

1. Introduction

All graphs considered will be finite, connected, undirected graphs without loops or multiple edges. A graph is defined to be distance-transitive if for all vertices u, v, x, y such that d(u, v) = d(x, y) there is an automorphism h of the graph such that uh = x, vh = y. We summarise the theory of distance-transitive graphs and describe various known existence results. We also show how one may attempt to extend some of these existence results to graphs of higher valency and mention some of the difficulties involved.

2. The theory of distance-transitive graphs

Let Γ be a distance-transitive graph of diameter d and valency k. Associate with Γ a $(d+1) \times (d+1)$ tridiagonal matrix

$$\mathbf{P}(\Gamma) = \begin{bmatrix} 0 & 1 & & & & & 0 \\ k & a_1 & c_2 & & & & \\ & b_1 & a_2 & \cdot & & & \\ & & b_2 & \cdot & \cdot & & \\ & & & \cdot & \cdot & & \\ & & & & \cdot & c_{d-1} & \\ & & & & & a_{d-1} & c_d \\ 0 & & & & & b_{d-1} & a_d \end{bmatrix}$$

where, if u and v are two vertices such that d(u, v) = i, and

$$\Gamma_i(u) = \{w \mid d(w, u) = i\},$$

$$c_i = |\Gamma_{i-1}(u) \cap \Gamma_1(v)|,$$

$$a_i = |\Gamma_i(u) \cap \Gamma_1(v)|,$$

$$b_i = |\Gamma_{i+1}(u) \cap \Gamma_1(v)| .$$

These numbers are independent of the choices of u and v. We call this matrix the <u>intersection matrix</u> of the graph. This is sometimes written in shorthand form

$$\left\{ \begin{matrix} * & 1 & c_2 & \cdot & \cdot & \cdot & \cdot & c_{d-1} & c_d \\ 0 & a_1 & a_2 & \cdot & \cdot & \cdot & \cdot & a_{d-1} & a_d \\ k & b_1 & b_2 & \cdot & \cdot & \cdot & \cdot & b_{d-1} & * \end{matrix} \right\}$$

and called the <u>intersection array.</u>

Example. If we draw the Petersen graph in the form

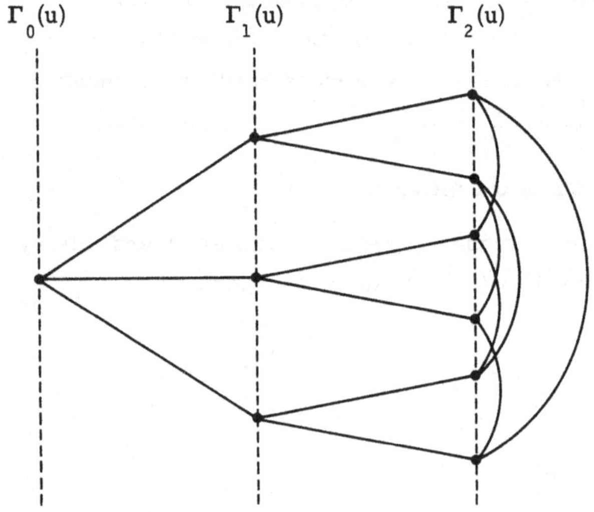

we have intersection array

$$\left\{ \begin{matrix} * & 1 & 1 \\ 0 & 0 & 2 \\ 3 & 2 & * \end{matrix} \right\} .$$

P(Γ) satisfies the following properties

(1) $c_i > 0,\ a_i \geq 0,\ b_i > 0,\ c_i + a_i + b_i = k\ \ (i = 1, 2, \ldots, d-1)$

$c_d + a_d = k$.

(2) If $k_i = |\Gamma_i(u)|$ then $k_0 = 1$, $k_1 = k$, $k_i = k_{i-1}b_{i-1}/c_i$ $(2 \leq i \leq d)$ and so the numbers k_i calculated from the matrix $\mathbf{P}(\Gamma)$ in this way are integers. We often write Γ_i for $\Gamma_i(u)$ when the choice of u is clear.

(3) $$1 \leq c_2 \leq c_3 \leq \ldots \leq c_d$$
$$k \geq b_1 \geq b_2 \geq \ldots \geq b_{d-1}.$$

A proof of this is given in [6].

(4) $\mathbf{P}(\Gamma)$ has $d + 1$ distinct real eigenvalues $k = \lambda_0 > \lambda_1 > \lambda_2 > \ldots > \lambda_d$. If $\mathbf{u}_i$ denotes the left eigenvector or $\mathbf{P}(\Gamma)$ with first entry 1 corresponding to λ_i, and $\mathbf{v}_i$ denotes the right eigenvector of $\mathbf{P}(\Gamma)$ with first entry 1 corresponding to λ_i, then the numbers $\frac{(\mathbf{u}_0, \mathbf{v}_0)}{(\mathbf{u}_i, \mathbf{v}_i)}$ are integers, where (,) denotes the usual inner product. For a proof of this see [1].

If G is the automorphism group of Γ we say Γ is primitive if G acts primitively on the vertices. Otherwise Γ is imprimitive.

Definition. We call a distance-transitive graph antipodal if for all $v_1, v_2 \in \Gamma_0(u) \cup \Gamma_d(u)$ either $d(v_1, v_2) = d$ or $v_1 = v_2$.

Theorem 1. A distance-transitive graph of valency > 2, diameter > 2 is imprimitive if and only if it is either bipartite or antipodal.

Proof. See [6].

Definition. Suppose Γ is distance-transitive and $\Gamma_0(u) \cup \Gamma_d(u)$ is a block of imprimitivity $(d > 2)$. Then we define the derived graph Γ' of Γ to be the graph with vertex set consisting of one vertex v_i for each block B_i in the system of imprimitivity determined by $\Gamma_0(u) \cup \Gamma_d(u)$, two vertices v_i, v_j of Γ' being joined if and only if some vertex $u_p \in B_i$ is adjacent to some vertex $u_q \in B_j$ in Γ.

Theorem 2. Let Γ be an antipodal graph of valency k and diameter $d > 2$. Then (1) the derived graph Γ' is distance-transitive,

(2) if d is the diameter of Γ then the diameter of Γ' is $[\frac{d}{2}]$.

Proof. See [6].

3. Known existence results

Several infinite families of distance-transitive graphs are known:

(1) The complete graphs K_n.

(2) The complete bipartite graphs $K_{n,n}$.

(3) The n-cubes Q_n.

(4) The graphs O_k whose vertices are labelled by unordered (k-1)-tuples of $2k - 1$ letters and the edges join every pair of vertices with no letter in common.

(5) The bipartite graphs $2.O_k$ with two sets of vertices, each labelled by unordered (k-1)-tuples of $2k - 1$ letters with two vertices joined if and only if they are in different sets of the bipartition and the labels have no letter in common.

A considerable amount is known about distance-transitive graphs of diameter 2. The reader is referred to [1] for a survey. Gardiner [5] has studied the existence of the graphs $n.K_{n,n}$ (antipodal graphs with $k_d = n - 1$ and derived graph $K_{n,n}$).

In addition the following complete classifications are known:

Theorem 3. The only trivalent distance-transitive graphs are those given in table 1.

Proof. See [4].

Theorem 4. The only 4-valent distance-transitive graphs are those given in table 2.

Proof. See [7], [8], [9].

The basic method employed in proving these two theorems is as follows:

(a) Find an upper bound for the order $|G_\alpha|$ of the stabiliser of a vertex. This is given in Theorem 3 by a result of Tutte and in Theorem 4 by results of Sims, Thompson, Gardiner and Quirin.

Table 1. The distance-transitive graphs of valency three

d	P(Γ)	$k_0, k_1, \ldots, k_d$	$\|V\|$	Name or reference
1	$\left\{\begin{matrix}*1\\02\\3*\end{matrix}\right\}$	1, 3	4	K_4
2	$\left\{\begin{matrix}*13\\000\\32*\end{matrix}\right\}$	1, 3, 2	6	$K_{3,3}$
2	$\left\{\begin{matrix}*11\\002\\32*\end{matrix}\right\}$	1, 3, 6	10	Petersen's graph
3	$\left\{\begin{matrix}*123\\0000\\321*\end{matrix}\right\}$	1, 3, 3, 1	8	Cube
3	$\left\{\begin{matrix}*113\\0000\\322*\end{matrix}\right\}$	1, 3, 6, 4	14	Heawood's graph
4	$\left\{\begin{matrix}*1123\\00000\\3221*\end{matrix}\right\}$	1, 3, 6, 6, 2	18	Pappus's graph
4	$\left\{\begin{matrix}*1112\\00011\\3221*\end{matrix}\right\}$	1, 3, 6, 12, 6	28	[2]
4	$\left\{\begin{matrix}*1113\\00000\\3222*\end{matrix}\right\}$	1, 3, 6, 12, 8	30	8-cage
5	$\left\{\begin{matrix}*11223\\000000\\32211*\end{matrix}\right\}$	1, 3, 6, 6, 3, 1	20	Desargues's graph
5	$\left\{\begin{matrix}*11123\\001100\\32111*\end{matrix}\right\}$	1, 3, 6, 6, 3, 1	20	Dodecahedron
7	$\left\{\begin{matrix}*1111113\\00001110\\3222111*\end{matrix}\right\}$	1, 3, 6, 12, 24, 24, 24, 8	102	[2]
8	$\left\{\begin{matrix}*11112223\\000000000\\32222111*\end{matrix}\right\}$	1, 3, 6, 12, 24, 24, 12, 6, 2	90	[6] - triple cover of 8-cage

(b) Use this upper bound for $|G_\alpha|$ to find an upper bound D for the diameter d.

(c) Use a computer to examine all tridiagonal matrices of size up to $(D+1) \times (D+1)$ with integer entries in the range $[0, k]$. The computer checks whether properties (1) to (4) of $\mathbf{P}(\Gamma)$ are satisfied. Matrices satisfying all these conditions are said to be feasible as the intersection matrix of a distance-transitive graph. In fact condition (4) is very restrictive and in these two cases there are very few feasible matrices.

(d) For each feasible matrix it remains to check whether or not it does correspond to a distance-transitive graph, and if it does, whether or not the graph is unique.

Theorems 3 and 4 have recently been proved by Gardiner without the use of a computer.

4. Extensions to higher valencies

Consider the various steps described above. Step (c) will work for graphs of any valency, although the amount of computer time required increases rapidly as the valency increases. It is essential even for small valencies to have the bound for the diameter as tight as possible.

The method employed for step (d) depends on the particular matrix, but one would hope to be able to identify any matrix of reasonably small valency and diameter.

Step (a) has been solved in the case of valency $p + 1$ (p prime) by A. Gardiner [12]. Gardiner proved that for an s-transitive graph of valency $p + 1$ (p prime), $s \in \{1, 2, 3, 4, 5, 7\}$, $|G_\alpha| \leq (p+1)p^{s-1}(p-1)^2$ $(s \in \{4, 5, 7\})$ and $|G_\alpha| \leq (p+1)!\,p!$ $(s \in \{1, 2, 3\})$. Since a distance-transitive graph is s-transitive for some s, this immediately gives a bound for $|G_\alpha|$.

Step (b) appears to be considerably more difficult in general than in the valency 3 or valency 4 cases. The author has recently found a method of bounding the diameter of a bipartite distance-transitive graph when a bound is known for $|G_\alpha|$ [10]. Combined with Gardiner's result this gives a bound for the diameter of a bipartite distance-transitive graph of valency $p + 1$. Assume the intersection array to be

Table 2. The distance-transitive graphs of valency four (L(Γ) denotes the line graph of Γ)

girth	d	P(Γ)	$k_0, k_1, \ldots, k_d$	|V|	Name or reference
3	1	$\begin{Bmatrix} *1 \\ 03 \\ 4* \end{Bmatrix}$	1, 4	5	K_5
3	2	$\begin{Bmatrix} *14 \\ 020 \\ 41* \end{Bmatrix}$	1, 4, 1	6	Octahedron
3	2	$\begin{Bmatrix} *12 \\ 012 \\ 42* \end{Bmatrix}$	1, 4, 4	9	The line graph of $K_{3,3}$ $L(K_{3,3})$
3	3	$\begin{Bmatrix} *114 \\ 0120 \\ 421* \end{Bmatrix}$	1, 4, 8, 2	15	L (Petersen's graph)
3	3	$\begin{Bmatrix} *112 \\ 0112 \\ 422* \end{Bmatrix}$	1, 4, 8, 8	21	L (Heawood's graph)
3	4	$\begin{Bmatrix} *1112 \\ 01112 \\ 4222* \end{Bmatrix}$	1, 4, 8, 16, 16	45	L (8-cage of valency 3)
4	2	$\begin{Bmatrix} *14 \\ 000 \\ 43* \end{Bmatrix}$	1, 4, 3	8	$K_{4,4}$
4	3	$\begin{Bmatrix} *134 \\ 0000 \\ 431* \end{Bmatrix}$	1, 4, 4, 1	10	[9]
4	3	$\begin{Bmatrix} *124 \\ 0000 \\ 432* \end{Bmatrix}$	1, 4, 6, 3	14	[9]
6	3	$\begin{Bmatrix} *114 \\ 0000 \\ 433* \end{Bmatrix}$	1, 4, 12, 9	26	6-cage
6	3	$\begin{Bmatrix} *112 \\ 0002 \\ 433* \end{Bmatrix}$	1, 4, 12, 18	35	O_4
4	4	$\begin{Bmatrix} *1234 \\ 00000 \\ 4321* \end{Bmatrix}$	1, 4, 6, 4, 1	16	4-cube Q_4
6	4	$\begin{Bmatrix} *1134 \\ 00000 \\ 4331* \end{Bmatrix}$	1, 4, 12, 12, 3	32	$4.K_{4,4}$ [11], [5]

12	6	$\begin{Bmatrix} *111114 \\ 0000000 \\ 433333* \end{Bmatrix}$	1, 4, 12, 36, 108, 324, 243	728	12-cage (may not be unique)
6	7	$\begin{Bmatrix} *1122334 \\ 00000000 \\ 4332211* \end{Bmatrix}$	1, 4, 12, 18, 18, 12, 4, 1	70	$2.O_4$

$$\begin{Bmatrix} * & 1 & c_2 & \cdots & c_q & \frac{k}{2} & \frac{k}{2} & \cdots\cdots & \frac{k}{2} & c_r & \cdots \\ 0 & 0 & 0 & \cdots & 0 & 0 & 0 & \cdots\cdots & 0 & 0 & \cdots \\ k & k-1 & b_2 & \cdots & b_q & \frac{k}{2} & \frac{k}{2} & \cdots\cdots & \frac{k}{2} & b_r & \cdots * \end{Bmatrix} \quad \begin{matrix} c_q < \frac{k}{2} \\ c_r > \frac{k}{2} \end{matrix}$$

Since $k_i > k_{i-1}$ $(1 \le i \le q+1)$ and $k_i < k_{i+1}$ $(r \le i \le d)$ and a bound for $|G_\alpha|$ implies a bound for $\max_i k_i$, we see that a bound for $|G_\alpha|$ implies a bound for q and a bound for $d - r$. It is shown in [10] that $r \le 3q + 2$ and so we can obtain a boundfor the diameter. Although the bound given is not good enough to make the computations in step (c) a practical proposition even for graphs of small valency, one would hope that the method in [10] is capable of considerable improvement.

The generalisation to the non-bipartite case is more difficult although it seems that there is some hope for a method similar to that used in [10]. Notice that if the graph is imprimitive but not bipartite it is antipodal [6] and has a primitive distance-transitive derived graph of the same valency and diameter $[\frac{d}{2}]$. Hence to complete step (b) it is sufficient to be able to convert a bound for $|G_\alpha|$ into a bound for the diameter of a primitive distance-transitive graph.

References

1. N. L. Biggs. Finite groups of automorphisms. London Mathematical Society Lecture Note Series 6, Cambridge University Press (1971).
2. N. L. Biggs. Three remarkable graphs. Canadian Journal of Mathematics, Vol. XXV, no. 2 (1973), 397-411.
3. N. L. Biggs. Intersection matrices for linear graphs. In Combinatorial Mathematics and its Applications, ed. D. J. A. Welsh, Academic Press (1971).

4. N. L. Biggs and D. H. Smith. On trivalent graphs. Bull. London Math. Soc., 3 (1971), 155-8.

5. A. Gardiner. Imprimitive distance-regular graphs and projective planes. To appear.

6. D. H. Smith. Primitive and imprimitive graphs. Quart. J. Math. Oxford (2) (1971), 551-7.

7. D. H. Smith. On tetravalent graphs. Journal of the London Mathematical Society, to appear.

8. D. H. Smith. Distance-transitive graphs of valency four. Journal of the London Mathematical Society, to appear.

9. D. H. Smith. On bipartite tetravalent graphs. To appear.

10. D. H. Smith. Bounding the diameter of a distance-transitive graph. J. Combinatorial Theory (B) 16 (1974), 139-44.

11. D. H. Smith. Highly symmetrical graphs of low valency. Ph. D. thesis, University of Southampton (1971).

12. A. Gardiner. Arc transitivity in graphs. Quart. J. Math., to appear.

Glamorgan Polytechnic,
Treforest, Pontypridd,
Glamorgan, Wales

ENUMERATION OF GRAPHS ON A LARGE PERIODIC LATTICE

H. N. V. TEMPERLEY

In statistical mechanics we are interested in various problems of counting subgraphs of a periodic lattice and I shall consider the plane square lattice as an illustration. Virtually all the analytically solved problems on this lattice either involve the counting of subgraphs all of whose vertices are of even valency, or can be transformed into such problems [3]. I consider a problem to be 'analytically solved' if we know the limiting form of the generating function in the limiting case when the numbers and lengths of the rows become very large.

One of the main methods of handling these problems is by a matrix method [3]. The matrix describes the operation of adding a row of points to the lattice. Each row of the matrix corresponded to one of the possible configurations of vertical lines in row m of the lattice, each column of the matrix to one of the possible configurations of vertical lines in row m + 1 of the lattice. Configurations of the whole lattice are then enumerated by a high power of this matrix and the crucial problem is to determine its largest eigenvalue. For a row of n points this is a $2^n \times 2^n$ matrix.

I have no brand new results to report, but I think that it will be useful to describe a method that has been used for quite a number of these problems. It was originally used by Bethe for the rather different problem of the magnetisation of a one-dimensional lattice. I think that it is helpful to give an illustration of it here, because it seems to have earned the unjust reputation of being abtruse and difficult. We restrict ourselves to a form of the so-called six-vertex problem, in which the only types of vertex allowed are

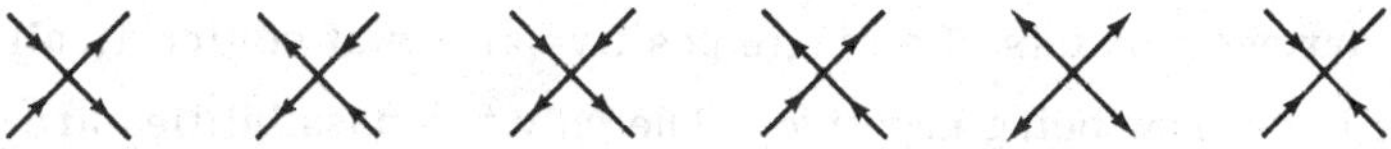

(1) (2) (3) (4) (7) (8)

Figure 1

We imagine the lattice to be built up by the addition of diagonal rows of points and lines and we reckon an arrow pointing to the north-east or south-east as positive, and implying the presence of the corresponding line in a subgraph, whereas arrows pointing to the north-west or south-west are reckoned as negative and imply the absence of the corresponding line in the subgraph. It is easily seen that all six types of vertex in Figure 1 have the following properties:

(a) Type (1) corresponds to the presence of four lines, arrows all positive, Type (2) to an isolated vertex, while types (3), (4), (7), (8) correspond to two lines or positive arrows at each vertex. Thus, in all cases we have an even number of lines. (The numbering of types of vertex in Figure 1 are equivalent to those in Figure 1 of [2].)

(b) As we move from one (diagonal) row of lines to the next, the number of positive arrows (lines present) is conserved.

For a vertex of type (1) there are two in each row.
For a vertex of type (2) there are 0 in each row.
For vertices of types (3), (4), (7), (8) there is one in each row.

Thus, restriction to the six types of vertex shown in Figure 1 greatly simplifies the structure of the transfer matrix. It falls into blocks, corresponding to 0, 1, 2, 3, ... positive arrows in any row, since a configuration of r positive arrows in one diagonal row necessarily implies that there are r positive arrows in the next diagonal row. We can label the arrows from left to right in successive rows. For an illustration, we consider the case when all six types of vertex are assigned equal weights in the generating function, all of the remaining ten possible dispositions or arrows being forbidden. This is equivalent to the celebrated 'square-ice' problem solved by Lieb [1], the restrictions on the arrow directions representing the chemical restrictions on the permitted positions of hydrogen nuclei in hydrogen bonded materials such as ice. (No more than two may be in the immediate neighbourhood of an oxygen atom.)

Consider first the elements of the transfer matrix if a diagonal row of arrows consists of a single positive arrow at station 1, all other arrows in the row being negative. The only two possibilities are a positive arrow at station 1 in the next row which corresponds to a vertex of type (8) in Figure 1, or at station 2 in the next row, corresponding to a vertex of type (4) in Figure 1. A positive arrow at station 2, all others in the row being negative, can correspond to a positive arrow at station 2 in the next row (vertex of type (7)) or to one at station 1 in the next row (vertex of type (3)). Thus, for the 'one positive arrow' problem the structure of the transfer matrix taking us from one diagonal row to the next is extremely simple and can be exhibited thus, the positive arrow never moving more than one step to the right or left as we move up one row:

$$\begin{aligned}(1) &\rightarrow (1) + (2)\\ (2) &\rightarrow (1) + (2)\\ (3) &\rightarrow (3) + (4)\\ (4) &\rightarrow (3) + (4)\end{aligned} \tag{1}$$

where, for example, (3) stands for a row configuration with a single positive arrow at station (3), all others in the row being negative. The eigenvalues of operation (1) are almost trivial to calculate. We replace (r) by α^r and find easily from (1) that the eigenvalues are $2 + \alpha + \alpha^{-1}$ and zero. If we have a periodic boundary condition, so that (r) and (r + n) are reckoned to be the same configuration, then α can be any of the nth roots of unity.

The situation is only a little more complicated for two positive arrows in each row. The 'single arrow rules' (1) apply independently to two arrows in stations such as (2) and (4) thus

$$(2, 4) \rightarrow (1, 3) + (1, 4) + (2, 3) + (2, 4) \tag{2}$$

the only exceptions being configurations of the type (1, 2), (3, 4) etc. for which the only possible vertex is one of type (1) in Figure 1 so that we have simply

$$\begin{aligned}(1, 2) &\rightarrow (1, 2)\\ (3, 4) &\rightarrow (3, 4) \text{ etc.}\end{aligned} \tag{3}$$

If it were not for the existence of the exceptional cases (3), we could solve for the eigenvalues of operation (2) simply by replacing (r, s) by $\alpha^r \ \beta^s$, whereupon we should calculate eigenvalues of the product type,

$$(2 + \alpha + \alpha^{-1})(2 + \beta + \beta^{-1}) \tag{4}$$

α and β being n^{th} roots of unity as before.

What I wish to point out is that the method in use for allowing for the exceptional configurations of the type (1, 2) is no more complicated than those in use for other wave-equation types of problem. We can satisfy both (2) and (3) by the assumption

$$(r, \ s) = \alpha^r \beta^s + P(\alpha, \ \beta)\alpha^s \beta^r. \tag{5}$$

Each term of (5) would satisfy (2) separately and we can find $P(\alpha, \ \beta)$, independent of r and s, such that (3) is satisfied also. If we define (r, s) and (s, r + n) as the same configuration we have

$$\alpha^{-n} = \beta^n = P(\alpha, \ \beta) \tag{6}$$

which means that α and β are determined by a transcendental equation instead of being n^{th} roots of unity. I would point out that, in physical systems e. g. in acoustics, quantum mechanics and elasticity, it is the exception, not the rule, for characteristic frequencies and wave-numbers to be harmonic.

This work generalises quite readily to three or more positive arrows per row. For three arrows we use a 'wave-function' of the type

$$(r, \ s, \ t) = \alpha^r \beta^s \gamma^t + P(\alpha, \ \beta)\beta^r \alpha^s \gamma^t + \dots \tag{7}$$

where there is one term with an appropriate numerical coefficient for each of the six permutations of α, β, γ.

Equation (6) is replaced by

$$\begin{aligned} \beta^n &= P(\alpha, \ \beta) \ P(\gamma, \ \beta) \\ \alpha^n &= P(\beta, \ \alpha) \ P(\gamma, \ \alpha) \\ \gamma^n &= P(\alpha, \ \gamma) \ P(\beta, \ \gamma) \ . \end{aligned} \tag{8}$$

For the special case that the various coefficients in (7) are just the signatures of the permutations, physicists will recognise the Slater determinant and pure mathematicians the determinants operated on by the third compound matrix of an operator like (1). Equations (8) are still transcendental equations of a relatively tractable type.

In the limiting case of large n, it turns out that the largest eigenvalue comes from $\frac{1}{2}n$ positive arrows per row, and the problem of determining α, β, γ, ... goes over into that of solving an integral equation which can be carried through in certain cases. I am suggesting that the problems are of the same basic type as in elasticity or quantum mechanics.

Temperley and Lieb [2] pointed out that the 'percolation' and 'chromatic polynomial' problems are, for the plane square lattice, reduced to problems of this type on an auxiliary plane square lattice 'staggered' in the sense that we have different six-vertex operators taking us between alternate pairs of diagonal rows. The method can in principle be generalised to deal with this situation, but only special cases have been solved so far.

References

1. D. H. Lieb. Phys. Rev., 162 (1967), 162-70.

2. H. N. V. Temperley and E. H. Lieb. Proc. Roy. Soc. A., 322 (1971), 151-80.

3. H. N. V. Temperley. The enumeration of graphs on large periodic lattices. In Combinatorics (ed. by D. J. A. Welsh and D. R. Woodall, Institute of Mathematics and its Applications, 1973).

University College,
Swansea, Wales

SOME POLYNOMIALS ASSOCIATED WITH GRAPHS

W. T. TUTTE

1. A recursion formula

Let us consider two familiar operations on a graph G with a specified link A, that of deleting A but retaining its end vertices, and that of contracting A, with its two ends, into a single vertex. Let the graphs obtained from G by these two operations be denoted by G'_A and G''_A respectively.

It is well known that the number $\mathcal{T}(G)$ of spanning trees of G, called the complexity or tree-number of G, satisfies the recursion formula

$$\mathcal{T}(G) = \mathcal{T}(G'_A) + \mathcal{T}(G''_A).$$

I came upon this formula at an early stage in my combinatorial career ([1]), and it inspired me to look for other functions of graphs with a similar recursion formula. I found several of them, the chromatic polynomial of G being already a well-known example ([3], footnote).

These other functions were multiplicative in the sense that the function of any graph G was equal to the product of the function over the components of G. In this respect they differed from the tree-number. So I became particularly interested in functions f satisfying the following two laws

(1) $f(G) = f(G'_A) + f(G''_A)$,

(2) $f(H + K) = f(H).f(K)$.

Here A is any link of G. The symbol H + K stands for a graph that is the union of two disjoint subgraphs H and K.

Perhaps the theory of the general function f is best described in terms of an infinite sequence $(X_0, X_1, X_2, \ldots)$ of special graphs

X_j. The graph X_j has a single vertex and exactly j edges, these being necessarily loops. We note the following rule.

(1.1) f(G) can be determined for an arbitrary graph G if $f(X_j)$ is known for each j.

Proof. If G has no link each of its components is an X_j, and we can determine f(G) by using (2). (For consistency we write f(G) = 1 when G is a null graph.) If G has a link we can use (1). So the theorem follows by induction over the number of edges.

We may now ask 'Is there an f corresponding to an arbitrary choice of the numbers $f(X_j)$?' The answer is 'Yes', and an explicit formula can be given in terms of the spanning subgraphs of G.

A spanning subgraph of G is determined by its set S of edges. We write the subgraph as G:S. Let N(S, m) denote the number of components of G:S with cycle-rank (or cyclomatic number) m. Let $(y_0, y_1, y_2, \ldots)$ be a sequence of indeterminates over the ring of 'numbers' being considered. Let us write

$$f(G) = \sum_S \left\{ \prod_{j=0}^{\infty} y_j^{N(S, j)} \right\}. \tag{3}$$

It is clear that f satisfies (2).

We can prove also that f satisfies (1), in the following way. First we partition the class of spanning subgraphs of G into two subclasses C_1 and C_2, according as A is or is not in S. Then the terms of the sum f(G) corresponding to the members of C_2 are the terms of the sum $f(G'_A)$. The contraction of A sets up a 1-1 correspondence between the spanning subgraphs (G:S) of G in C_1 and the spanning subgraphs $(G:S)''_A$ of G''_A. The graphs G:S and $(G:S)''_A$ have the same components, except for the replacement of the component H of G:S containing A by H''_A. Moreover it is easily verified that H and H''_A have the same cycle-rank. Accordingly the terms of the sum f(G) corresponding to members of C_1 are the terms of the sum $f(G''_A)$. We thus have

(1.2) The function f(G) defined by (3) satisfies the basic recursion formulae (1) and (2).

Let us apply (3) to the graph $G = X_j$. We have

$$f(X_j) = \sum_{i=0}^{j} \binom{j}{i} y_j \quad (j = 0, 1, 2, \ldots). \tag{4}$$

These equations can be inverted as

$$y_j = \sum_{i=0}^{j} (-1)^i \binom{j}{i} f(X_i). \tag{5}$$

It follows that we can assign values to the y_j so as to make f take arbitrarily assigned values for the graphs X_j. So by (1.1), Formula (3) gives the most general function f satisfying (1) and (2). In the terminology of ([2]) it gives the most general V-function.

We now consider some examples. If y_j is set equal to 1 for all j then f(G) is the number of spanning subgraphs of G, by (3). Hence

$$f(G) = 2^{\alpha_1(G)},$$

where $\alpha_1(G)$ is the number of edges of G.

If we put $f(X_j) = 1$ for each j then $y_j = 1$ if $j = 0$, and $y_j = 0$ if $j > 0$. By (3), f(G) is then the number of spanning subgraphs of G that are forests.

If $P(G, \lambda)$ is the chromatic polynomial of G we have the recursions

$$P(H + K, \lambda) = P(H, \lambda)P(K, \lambda),$$
$$P(G, \lambda) = P(G'_A, \lambda) - P(G''_A, \lambda),$$

where A is any link of G. It follows that the function

$$f(G) = (-1)^{\alpha_0(G)} P(G, \lambda),$$

where $\alpha_0(G)$ is the number of vertices of G, is a V-function. It satisfies $f(X_0) = -\lambda$, and $f(X_j) = 0$ if $j > 0$.

2. Trivalent graphs

Consider a divalent vertex b of G, incident with two links A_1 and A_2. Let their other ends be a_1 and a_2 respectively. We can obtain a new graph H from G by deleting b, A_1 and A_2, and then adjoining a new edge A with ends a_1 and a_2. It may happen that a_1 and a_2 coincide, in which case A is a loop of H. We say that H is derived from G by suppressing b. If f is a V-function such that $f(H) = f(G)$ whenever H is derived from G by the suppression of a divalent vertex, then we say that f is topologically invariant. A trivial example is the zero V-function, taking the value 0 for every non-null graph.

(2.1) Let f be any non-zero V-function. Then f is topologically invariant if and only if $f(X_0) = -1$.

Proof. We use the above notation, writing also K and L for the graphs derived from G and H by deleting A_1 and A respectively. Evidently

$$\begin{aligned} f(G) - f(H) &= f(K), \quad \text{by (1)} \\ &= f(L + X_0) + f(L), \quad \text{by (1)} \\ &= f(L).\{f(X_0) + 1\}, \quad \text{by (2)}. \end{aligned}$$

If $f(X_0) = -1$ we deduce that in every case $f(G) = f(H)$. Conversely if f is known to be topologically invariant we choose G so that $f(L) \neq 0$ and deduce that $f(X_0) = -1$.

A minor question of terminology arises here. Shall we say that the divalent vertex in X_1 can be suppressed? This case is not covered by the above discussion; there the divalent vertex had to be incident with two links. We say here that the vertex in X_1 can be suppressed and that X_1 is then converted into a graph with no vertices and one edge. Such a 'graph' is conveniently represented by a simple closed curve in a diagram. Let us call it a vortex. A topologically invariant V-function is to be taken as having the same value for a vortex as for X_1.

A graph is trivalent if each vertex is incident either with three links or with one link and one loop.

Let A be an edge of a trivalent graph G. Let its ends be x and y. Let the other edges incident with x be A_1 and A_2, and let the edges other than A incident with y be A_3 and A_4. The edges A_1,

A_2, A_3 and A_4 need not all be distinct. Thus A_1 and A_2 may coincide as a loop incident with x, or A_1 and A_3 may coincide as a second link joining x and y.

To twist A in G is to alter the incidences of x and y in the following way. x is taken to be incident with A, A_1 and A_3 instead of A, A_1 and A_2, and y with A, A_2 and A_4. Let us write G_1 for the resulting graph, in which the incidence relations of the other vertices are unaltered. The typical cases are indicated in Figure 1.

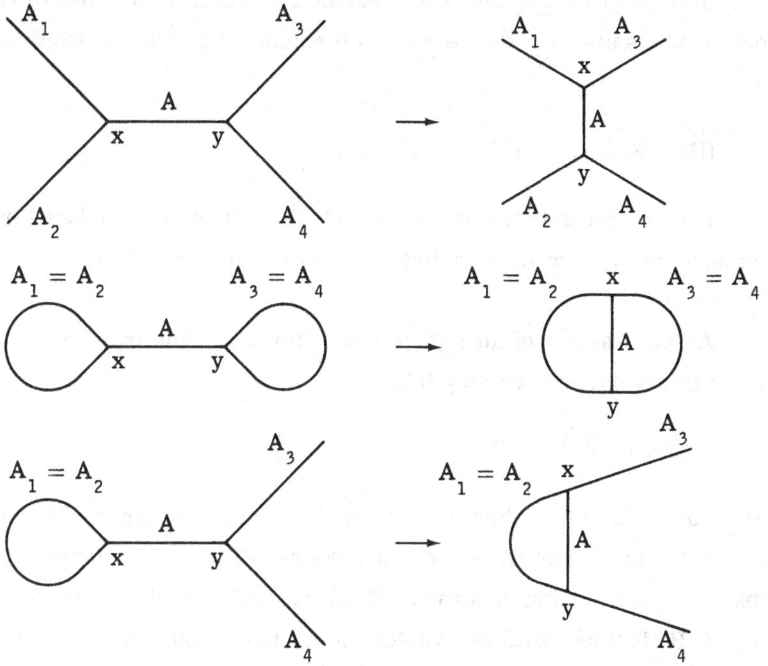

Figure 1

Actually an edge A can, in general, be twisted in two distinct ways since the suffixes of A_3 and A_4 can be interchanged. The effect of any twist can be annulled by another twist of the same edge.

We observe that, starting with either G or G_1, the same graph is obtained by contracting A. Hence, by (1), any V-function satisfies

$$f(G) - f(G'_A) = f(G_1) - f((G_1)'_A).$$

Let us denote by G' the result of suppressing the divalent vertices x and y in G'_A, and by G'_1 the result of suppressing them in $(G_1)'_A$. Then if f is topologically invariant we have

$$(6) \qquad f(G) - f(G_1) = f(G') - f(G'_1).$$

This is of interest as being an identity involving trivalent graphs only. Vortices are to be counted as trivalent.

In ([2]) an F-function is defined as a function f of non-null trivalent graphs satisfying the recursion formula (6) and the multiplicative rule

$$(7) \qquad f(H + K) = f(H).f(K).$$

It is shown in that paper that each F-function f is induced by some uniquely determined topologically invariant V-function g satisfying $g(X_0) = -1$.

As an example of an F-function f for which the inducing V-function is not obvious we may take

$$(8) \qquad f(G) = (-1)^{n(G)}Q(G),$$

where $2n(G)$ is the number of vertices of the trivalent graph G, and $Q(G)$ is the number of its 2-factors. We recall that an m-factor of a graph K is a spanning subgraph R of K such that the number of edges of R incident with any vertex, loops being counted twice, is m. Thus $Q(G)$ is also the number of 1-factors of G.

We leave it to the reader to relate the 2-factors of G, G_1, G' and G'_1 and thus to verify that f is indeed an F-function. He can of course refer to ([2]). The number of spanning subgraphs of a vortex is 2, one being null, and each of them is a 2-factor.

The F-function of Formula (8) has a polynomial generalization. Let $\Pi_k(G)$ denote the number of spanning subgraphs of the trivalent graph G that are without isolated vertices and that have exactly $2k$ vertices of odd valency, that is incident with an odd number of links.

Write

$$(9) \qquad D(G, x) = (-1)^{n(G)} \sum_{k=0}^{n(G)} \Pi_k(G)x^k ,$$

where x is an indeterminate. Then it can be verified, as in ([2]), that the polynomial D(G, x) is an F-function.

The value of $g(X_j)$ for the corresponding V-function g is still undetermined for general j. However for the F-function of Equation (8) the corresponding V-function g satisfies

$$(10) \qquad g(X_j) = \tfrac{1}{2}(-1)^j \{3^j + 1\}.$$

References

1. R. L. Brooks, A. H. Stone, C. A. B. Smith and W. T. Tutte. The dissection of rectangles into squares. Duke Math. J., 7 (1940), 312-40.
2. W. T. Tutte. A ring in graph theory. Proc. Cambridge Philos. Soc., 43 (1947), 26-40.
3. Hassler Whitney. The coloring of graphs. Ann. Math., 33 (1932), 688-718.

EQUIDISTANT POINT SETS

J. H. VAN LINT

In this talk we shall consider two problems which are both concerned with a set S of points in a metric space (R, d) such that for any 2 distinct points of S the distance d(x, y) is the same. Both problems are connected to several areas of combinatorial theory in the sense that these areas provide examples which often turn out to meet certain bounds which one can derive for these equidistant sets. One other analogy seems to be the fact that we do not really understand these problems yet.

1. Equiangular lines

In our first problem we take R to be elliptic space of dimension r - 1 and d to be elliptic distance. It is more convenient to describe this space by considering the lines through the origin in r-dimensional euclidean space $\mathbf{R}^r$ and defining the distance to be the angle between two such lines.

Definition. (i) $v_\alpha(r)$ is the maximum number of lines in $\mathbf{R}^r$ such that each pair of these lines makes an angle $\arccos \alpha$, $\alpha > 0$.
(ii) $v(r) := \max\{v_\alpha(r) \mid 0 < \alpha \leq 1\}$.

In 1965 Van Lint and Seidel [5] treated this problem for $r \leq 7$. A few months ago a paper by Lemmens and Seidel [3] appeared which extended the results to $r \leq 43$, however with a number of gaps. E. g. the value of v(14) is not known. These results revived my own interest in the problem. It seems worthwhile to point out some of the interesting connections to other areas of combinatorial theory. For a survey of the present state of affairs concerning v(r) we refer to [3].

Let S be a set of v unit vectors spanning $\mathbf{R}^r$ such that any two distinct vectors in S have inner product $\pm\alpha$. If G is the Gram matrix of S, then we write $A := \alpha^{-1}(G - I)$. Then A is a symmetric matrix with zero diagonal and all other entries ± 1. Since G has smallest eigenvalue 0 with multiplicity $v - r$ the smallest eigenvalue of A is $-\alpha^{-1}$ with multiplicity $v - r$. In this way the problem of finding equidistant point sets is reduced to finding such $(0, +1, -1)$-matrices A such that the smallest eigenvalue is ≤ -1 and has a large multiplicity. Any such $(0, +1, -1)$-matrix A can be interpreted as the adjacency matrix of a graph on vertices $P_1, P_2, \ldots, P_v$ by including the edge $\{P_i, P_j\}$ iff $a_{ij} = -1$. Many good examples are connected to strong graphs which we now define. (We exclude void and complete graphs.)

Definition. Let A be the $(0, +1, -1)$-adjacency matrix of a graph on the vertices $P_1, P_2, \ldots, P_v$. If there are two integers p_1 and p_2 such that for any two vertices P_i, P_j with $a_{ij} = (-1)^h$ there are exactly p_h points joined by an edge to one, but not both, of P_i and P_j, then the graph is called <u>strong</u>. If the graph is also regular it is called <u>strongly regular</u>.

The following theorem makes it clear why these graphs are interesting for our problem.

Theorem 1 (cf. e.g. [6]). <u>A nonvoid and noncomplete graph of order</u> v <u>is strong if and only if its</u> $(0, +1, -1)$-<u>adjacency matrix satisfies</u>

$$(A - \rho_1 I)(A - \rho_2 I) = (v - 1 + \rho_1\rho_2)J, \quad (\rho_1 > \rho_2).$$

Clearly A has at most 3 distinct eigenvalues.

In [5] it was shown that $v(5) = 10$. The example was provided by the well known Petersen graph (five eigenvalues -3, five eigenvalues $+3$).

The following theorem shows how combinatorial designs can be combined to construct equidistant point sets.

Theorem 2. <u>If the projective plane</u> PG(2, q) <u>exists and if a Hadamard matrix of order</u> $q + 2$ <u>exists, then</u> $v(q^2+q+1) \geq (q+2)(q^2+q+1)$.

Proof. Let B be the incidence matrix of the plane. Let H be the Hadamard matrix with the first column consisting of 1's only. Delete the first column to obtain H_0. We replace each row of B by $q + 2$ new rows obtained by leaving the 0's where they are and replacing the 1's by the rows of H_0. We thus obtain a matrix A with $(q+2)(q^2+q+1)$ rows and $q^2 + q + 1$ columns such that any 2 rows have inner product ± 1 and every row is a vector of length $(q+1)^{\frac{1}{2}}$ in $\mathbf{R}^{q^2+q+1}$.

Due to our poor knowledge of projective planes the only example presently known whose order satisfies the conditions of Theorem 2 is $q = 2$ which yields $v(7) \geq 28$ (cf. [5]). Of course we can prove a more general theorem by taking B to be the incidence matrix of any block design with $\lambda = 1$ but this never gives examples near to known bounds. However, the incidence matrix of $PG(2, 2^l)$ in which we replace each line by a Hadamard matrix extended with a column of 1's yields the bound

$$v(2^{2l} + 2^l + 1) \geq 2^l(2^{2l} + 2^l + 1),$$

which is close to the result of Theorem 2.

Recently D. E. Taylor [7] proved that the inequality of Theorem 2 holds if $q + 1$ is a power of an odd prime. The examples obtained by this construction are best possible for small values of the parameter. We describe his construction. Let $q = p^n$ $(p \neq 2)$, (not the same q as above). Let $K = GF(q^2)$, V the 3-dimensional vector space over K and $\mathcal{P}(V)$ the corresponding projective plane. The equation

$$F(x_1, x_2, x_3; y_1, y_2, y_3) = x_1y_3^q + x_2y_2^q + x_3y_1^q = 0$$

defines a unitary polarity of $\mathcal{P}(V)$ (cf. [2]). Let $\mathcal{U}$ be the associated unital (absolute points, nonabsolute lines). Then $\mathcal{U}$ has $q^3 + 1$ points ([2], exercise 2.41). Take the line with equation $x_1 = 0$ as line at infinity and let ∞ be the point $(0, 0, 1)$ of $\mathcal{U}$. Then the q^3 other points of $\mathcal{U}$ are described by affine coordinates x, y. On these $q^3 + 1$ points we define a graph G as follows:

(1) (x, y) is joined to (a, b) if $\begin{cases} F(1, x, y; 1, a, b) \text{ is a square, } q \equiv -1 \pmod 4, \\ F(1, x, y; 1, a, b) \text{ is 0 or a non-square, } q \equiv 1 \pmod 4. \end{cases}$

(ii) ∞ is joined to all other points of $\mathcal{U}$.

Then G turns out to be a strong graph with incidence matrix A satisfying

$(A + qI)(A - q^2 I) = 0.$

Consequently, we have the following theorem.

Theorem 3. If $q = p^n$, $p \neq 2$, p prime, then

$v(q^2 - q + 1) \geq q^3 + 1.$

Since for $\varepsilon > 0$ and r sufficiently large there is a prime power between r and $r(1 + \varepsilon)$ we have

Theorem 4. $\lim_{r \to \infty} r^{-3/2} v(r) \geq 1.$

It is not difficult to show (cf. [3], [7]) that $v(r) \leq \frac{1}{2} r(r + 1)$ but I conjecture that actually $r^{3/2}$ gives the correct order of growth of $v(r)$.

2. Equidistant codes

Now let R be n-dimensional vector space over $GF(q)$ and let d be Hamming distance, defined by $d(\underline{x}, \underline{y}) := |\{i \mid x_i \neq y_i\}|$. We define the weight $w(\underline{x})$ of $\underline{x}$ by $w(\underline{x}) := d(\underline{x}, \underline{0})$.

Definition. An equidistant (m, k)-code is an m-subset S of R such that

$$\forall \underline{x} \in S \, \forall \underline{y} \in S \, [\underline{x} \neq \underline{y} \Rightarrow d(\underline{x}, \underline{y}) = k].$$

If H is a Hadamard matrix of order n then the n rows of $\frac{1}{2}(H + J)$ form an equidistant binary $(n, \frac{1}{2}n)$-code. From now on we take $q = 2$. With an equidistant $(m, 2k)$-code S we associate the matrix C which has as its rows all the code words of S. Each column C of S, interpreted as a binary vector, has a weight. If all these weights are 0, 1, $m - 1$ or m, we call S a trivial equidistant code. E. g. if $C = (I_m I_m \ldots I_m)$, k copies of I_m, then S is trivial with distance $2k$.

Let B be the incidence matrix of $PG(2, k)$ and let J be the $k^2 + k + 1$ by $k - 1$ matrix of 1's. Then

$$C = \begin{pmatrix} 0 & 0 & \ldots & 0 & \vdots & 0 & \ldots & 0 \\ & & B & & \vdots & & J & \end{pmatrix}$$

represents an equidistant $(k^2 + k + 2, 2k)$-code which is nontrivial. It was shown by M. Deza [1] that a nontrivial equidistant $(m, 2k)$-code has $m \le k^2 + k + 2$. We now announce the following theorem ([4]).

Theorem 5. _If a nontrivial equidistant_ $(k^2 + k + 2, 2k)$-_code exists, then the projective plane_ $PG(2, k)$ _exists._

Proof. We present a proof here which is shorter than the original proof given in [4]. We first remark that we can choose any row in C and by interchanging 0's and 1's change this row into a row of 0's. Then the other rows of the new matrix C all have weight 2k and any two of them have inner product k. In the following we always assume that C has a 0-row. We again use m for the number of rows of C. If a column of C has weight t then without loss of generality we can take this to be the first column and assume its t 1's are at the top. Let α_i be the number of 1's in the first t positions of column i and let β_i be the number of 1's in the last $m - t$ positions of column i. We define $a_i := \alpha_i/t$, $b_i := \beta_i/(m - t)$. Now we calculate the sum of the distances between respectively the first t rows, the last $(m - t)$ rows and between these two sets. We find

$$\sum a_i(1 - a_i) = k - k/t$$

$$\sum b_i(1 - b_i) = k - k/(m - t)$$

$$\sum \{a_i(1 - b_i) + b_i(1 - a_i)\} = 2k - 1.$$

Hence we have

$$\sum (a_i - b_i)^2 = -1 + \frac{k}{t} + \frac{k}{m - t},$$

i.e.

$$t(m - t) \le mk.$$

Substituting $m = k^2 + k + 2$ we see that $t \le k + 1$ or $t \ge k^2 + 1$. In the first case we call the column _light_, in the second case we call it _heavy_.

Now suppose there were $k + 2$ heavy columns. If any row had $k + 2$ 1's in these $k + 2$ positions all the others would have at most k

1's in these positions. In the same way there can be no more than $k + 2$ rows having $k + 1$ 1's in these $k + 2$ positions. In both cases the m rows together cannot have $(k + 2)(k^2 + 1)$ 1's in these $k + 2$ positions which contradicts the fact that the columns are heavy.

Now consider any row having q of its 1's in heavy columns. Clearly the sum of the inner products of the other rows with this row is at most $q(k^2 + k) + (2k - q)k$. Since this sum equals $k(k^2 + k)$ we have now shown that there are $k - 1$, k or $k + 1$ heavy columns.

If there are $k + 1$ heavy columns then by the reasoning used above there is a row having 1's in these $k + 1$ positions. Changing this row into the 0-row we find a C with only $k - 1$ heavy columns. If there are k heavy columns and the code is not trivial then there is a row having only $k - 1$ 1's in these heavy columns. Changing this row into the 0-row we find a C with $k + 2$ heavy columns, a contradiction. Therefore, if the equidistant code exists it can be represented by a C with the form

$$C = \begin{pmatrix} 0 & \cdots & 0 & \vdots & 0 & \cdots & 0 \\ & J & & \vdots & & B & \end{pmatrix}$$

where J has $k - 1$ columns. Now each row of B has $k + 1$ 1's and any two distinct rows of B have exactly one 1 in common. Since every column of B has at most $k + 1$ 1's, every column of B must have exactly $k + 1$ 1's. Hence B is the incidence matrix of $PG(2, k)$. This completes the proof.

In the cases where $PG(2, k)$ does not exist, e. g. $k = 6$, we have not been able to find the maximum number of code words in an equidistant code. For $k = 6$ this number is at least 32 since $PG(2, 5)$ exists and at most 43 by Theorem 5. Since $EG(2, 6)$ does not exist we tried to show that an equidistant $(m, 12)$-code has $m < 37$. By the same methods as used in the proof of Theorem 5 we could show that the existence of an equidistant binary $(37, 12)$-code implies the existence of an equidistant $(29, 6)$-code of word length 7 over an alphabet of 6 symbols which seems very unlikely. The work is being continued.

An obvious thing to try when one is looking for equidistant codes is to let C have the same form as above, i. e.

$$C = \begin{pmatrix} C & 0 & \dots & 0 & \vdots & 0 & 0 & \dots & 0 \\ & & B & & \vdots & & & J & \end{pmatrix}$$

where B is the v by b point-block incidence matrix of a block design with parameters $(v, k; b, r, \lambda)$ and J is the v by $r - 2\lambda$ matrix of 1's. Then C represents an equidistant $(v + 1, 2(r - \lambda))$-code. That this code cannot have many words is shown as follows.

Let $r - \lambda = d$. From the necessary conditions for v, k, b, r and λ we find (taking $k \le \frac{1}{2}v$, w. l. o. g.)

$$d = \frac{\lambda(v-1)}{k-1} - \lambda = \frac{\lambda(v-k)}{k-1},$$

i. e.

$$v = \frac{d(k-1)}{\lambda} + k \le \frac{d(r-1)}{\lambda} + r = \frac{d(d+\lambda-1)}{\lambda} + d + \lambda \le d^2 + d + 1,$$

where the last inequality is very weak unless $\lambda = 1$. For instance if $d = 6$, then an example yielding more than 32 words would have to have $v \ge 32$, hence $k \ge 6$, i. e. it would be EG(2, 6) which does not exist. It seems likely that a nontrivial equidistant (m, 12)-code has $m \le 32$.

References

1. M. Deza. Une propriété extremale des plans projectifs finis dans une classe de codes equidistants. Discrete Math. 6 (1973), 343-52.
2. D. R. Hughes and F. C. Piper. Projective planes. Springer Verlag, New York etc. (1973).
3. P. W. H. Lemmens and J. J. Seidel. Equiangular lines. Journal of Algebra, 24 (1973), 494-512.
4. J. H. van Lint. A theorem on equidistant codes. Discrete Math. 6 (1973), 353-8.
5. J. H. van Lint and J. J. Seidel. Equilateral point sets in elliptic geometry. Proc. Kon. Ned. Akad. v. Wetensch. Ser. A, 69 (1966), 335-48.
6. J. J. Seidel. Strongly regular graphs with (-1, 1, 0) adjacency matrix having eigenvalue 3. Linear Algebra and its Applications, 1 (1968), 281-98.

7. D. E. Taylor. Some topics in the theory of finite groups. Ph. D. thesis, Univ. of Oxford (1971).

Technological University,
Eindhoven, Netherlands

EIGENVALUES OF GRAPHS AND OPERATIONS

D. A. WALLER

Introduction

By the eigenvalues of the graph G is meant the eigenvalues of the adjacency matrix $A(G) = [a_{ij}]$ of G. If G has vertex-set $\{V_1, \ldots, V_m\}$ then its adjacency matrix has entries a_{ij} equal to the number of edges from V_i to V_j. We restrict attention to graphs without multiple edges between distinct vertices, but allow finite numbers of loops at the vertices.

The eigenvalues of a graph contain geometrical information (see for example Wilson's survey [4]). Graphs admit various operations (unary, binary, n-ary) with geometrical interpretation, and we investigate problems concerning the effect of such operations on the eigenvalues of the graphs involved. We conclude with an application of this work to the enumeration of spanning trees in graphs.

1. n-ary operations: disjoint union and n-fold join

Let $\phi(G)$ denote the characteristic polynomial of the graph G, and $\amalg$ the n-ary operation of disjoint union. One easily obtains:

1.1 Proposition. $\phi(\coprod_{j=1}^{n} G_j) = \prod_{j=1}^{n} \phi(G_j)$.

The join of n graphs $G_1, \ldots, G_n$ is the 1-dimensional skeleton of their join as simplicial complexes. Thus if the graph G_j has vertex-set $\{V_1^j, \ldots, V_{m_j}^j\}$, $j = 1, \ldots, n$, then the join $\mathop{*}_{j=1}^{n} (G_j) = G_1 * G_2 * \ldots * G_n$ consists of all the G_j as mutually disjoint subgraphs, together with an edge $e(V_r^j, V_s^k)$ for each pair of vertices $V_r^j \in G_j$, $V_s^k \in G_k$, $j \neq k$.

The main aim in this section is to relate the eigenvalues of $\mathop{*}_{j=1}^{n} (G_j)$ to those of the constituent G_j's for arbitrary regular graphs G_j. In fact regular graphs behave nicely under joining, to a large extent because they each have an adjacency matrix whose row-sums are all equal. The latter (algebraic) form of regularity should not be confused with (geometric) regularity (each vertex being joined by an edge to the same number of other vertices). To clarify this distinction, we give:

1.2 Definition. A graph is called row-regular (of valency p) if its adjacency matrix has all its row-sums equal to p.

All regular graphs are row-regular. Many graphs involving loops, for example

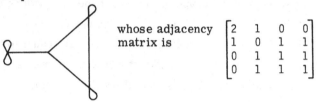

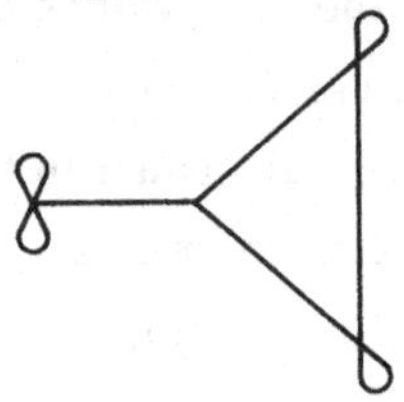

whose adjacency matrix is $\begin{bmatrix} 2 & 1 & 0 & 0 \\ 1 & 0 & 1 & 1 \\ 0 & 1 & 1 & 1 \\ 0 & 1 & 1 & 1 \end{bmatrix}$

are row-regular, but not (geometrically) regular. The importance of such graphs is demonstrated in [3]. Note that the number of loops is equal to the sum of the eigenvalues.

The Perron-Frobenius Theorem applies (essentially as used in [4]) to row-regular graphs to give:

1.3 Theorem. (i) If G is row-regular of valency p, then its largest eigenvalue is p.

(ii) If also G is connected then p is a simple root of $\phi(G)$, with all associated eigenvectors of the form $\boldsymbol{\alpha} = [\alpha, \ldots, \alpha]^t$.

(iii) If G is disconnected (equal to $\coprod_{j=1}^{r} G_j$), then the multiplicity of p is equal to the number r of components of G. The eigenvalue p corresponding to the jth component of G has eigenvectors of the form $[0 \ldots 0\ \boldsymbol{\alpha}\ 0 \ldots 0]^t$. (Notation is abused in such expressions, where vectors or matrices are built from smaller ones. Here **0** denotes any zero vector, and the jth 'subvector' is $\boldsymbol{\alpha}$.)

(iv) If $[\alpha_1, \ldots, \alpha_n]^t$ is an eigenvector for $\lambda < p$, then $\sum_{q=1}^{n} \alpha_q = 0$.

Although binary join is associative, one cannot proceed to the eigenvalues of n-fold joins of arbitrary regular graphs by induction, as the join of two regular graphs may not be regular (and joins involving irregular graphs do not have nicely related eigenvalues). However n-fold joins can be dealt with directly:

1.4 Theorem. *Let* $G_1, \ldots, G_n$ *be row-regular graphs,* G_j *having* m_j *vertices of valency* p_j*, and eigenvalues* $p_j, \lambda^j_2, \ldots, \lambda^j_{m_j}$*. Then*

(i) *each of the eigenvalues* $\lambda^j_2, \ldots, \lambda^j_{m_j}$, $j = 1, \ldots, n$ *is an eigenvalue of* $\overset{n}{\underset{j=1}{*}} (G_j)$;

(ii) *the other* n *eigenvalues of* $\overset{n}{\underset{j=1}{*}} (G_j)$ *are the eigenvalues of the matrix*

$$\begin{bmatrix} p_1 & m_2 & \cdot & \cdot & \cdot & m_n \\ m_1 & p_2 & \cdot & \cdot & & \cdot \\ \cdot & \cdot & \cdot & & \cdot & \cdot \\ \cdot & & \cdot & \cdot & \cdot & m_n \\ \cdot & & & \cdot & \cdot & \\ m_1 & \cdot & \cdot & \cdot & m_{n-1} & p_n \end{bmatrix}$$

Proof. (a) *The case* G_j *connected.*

(i) It follows from 1.2 (ii), (iv) that if λ is one of the eigenvalues denoted λ^j_i, $i = 2, \ldots, m_j$, then $\lambda < p_j$, and λ has associated eigenvectors of the form $\boldsymbol{\alpha}^j = [\alpha_1, \ldots, \alpha_{m_j}]^t$, with $\sum_{q=1}^{m_j} \alpha_q = 0$. Then

$$\begin{bmatrix} A(G_1) & 1 & \cdot & \cdot & \cdot & \cdot & 1 \\ 1 & A(G_2) & & & & & \cdot \\ \cdot & \cdot & \cdot & & & & \cdot \\ \cdot & & \cdot & \cdot & & & \cdot \\ \cdot & & & \cdot & \cdot & & \cdot \\ \cdot & & & & \cdot & \cdot & 1 \\ 1 & \cdot & \cdot & \cdot & \cdot & 1 & A(G_n) \end{bmatrix} \begin{bmatrix} 0 \\ 0 \\ \vdots \\ 0 \\ \boldsymbol{\alpha}^j \\ 0 \\ \vdots \\ 0 \end{bmatrix} = \begin{bmatrix} \Sigma\alpha^j_q \\ \vdots \\ \Sigma\alpha^j_q \\ \lambda\boldsymbol{\alpha}^j \\ \Sigma\alpha^j_q \\ \vdots \\ \Sigma\alpha^j_q \end{bmatrix} = \lambda \begin{bmatrix} 0 \\ \vdots \\ 0 \\ \boldsymbol{\alpha}^j \\ 0 \\ \vdots \\ 0 \end{bmatrix}$$

where the rth of the 0's in each of the column vectors is an abbreviation for 0_{m_r}, $r = 1, \ldots, j-1, j+1, \ldots, n$. Thus λ is an eigenvalue for $\underset{j=1}{\overset{n}{*}} (G_j)$, with associated eigenvectors of the form $[0 \ldots 0\ \boldsymbol{\alpha}^j\ 0 \ldots 0]^t$.

(ii) Next we consider the largest eigenvalue p_j of G_j, equal to the common valency of G_j. Since G_j is connected, the eigenvectors of p_j are of the form $\boldsymbol{\alpha}^j = [\alpha^j, \ldots, \alpha^j]^t$ i. e. all components are equal. Thus

$$[A(G_j)][\alpha^j, \ldots, \alpha^j]^t = p_j[\alpha^j, \ldots, \alpha^j]^t, \quad j = 1, \ldots, n.$$

Suppose μ is an eigenvalue for the join, with eigenvector $[\boldsymbol{\alpha}^1\ \boldsymbol{\alpha}^2 \ldots \boldsymbol{\alpha}^n]^t$, ($\boldsymbol{\alpha}^i$ denoting a vector with m_i components). Then the eigenvector equation $(\mathbf{M} - \mu I)[\boldsymbol{\alpha}^1 \ldots \boldsymbol{\alpha}^n]^t = 0$ gives $M = \sum_{j=1}^{n} m_j$ linear equations:

$$\begin{cases} (p_1 - \mu)\alpha^1 + m_2\alpha^2 + \cdots + m_n\alpha^n = 0 \\ m_1\alpha^1 + (p_2-\mu)\alpha^2 + \cdots + m_n\alpha^n = 0 \\ \vdots \\ m_1\alpha^1 + m_2\alpha^2 + \ldots + m_{n-1}\alpha^{n-1} + (p_n-\mu)\alpha^n = 0 \end{cases}$$

i. e. n <u>distinct</u> equations, the jth of these M equations being repeated m_j times, $j = 1, \ldots, n$. It follows that μ is a solution of this system if and only if it is one of the n eigenvalues of the matrix given in the statement of the theorem.

If G_j is connected for all j, then (i) provides $\sum_{j=1}^{n} (m_j - 1)$ and (ii) provides n eigenvalues, giving in all the M eigenvalues.

(b) <u>The case G_j disconnected.</u>

If G_j has k components, then its largest eigenvalue p_j has multiplicity k, and eigenspace E_j of dimension k spanned by $[\boldsymbol{\alpha}\ 0 \ldots 0]^t, [0\ \boldsymbol{\alpha}\ 0 \ldots 0]^t, \ldots, [0 \ldots 0\ \boldsymbol{\alpha}]^t$. E_j splits into a direct su $U \oplus W$ where U is spanned by the eigenvector $[\boldsymbol{\alpha} \ldots \boldsymbol{\alpha}]^t$ acting as in (ii) above, and W is spanned by

$$[\tfrac{1}{n_1}(1), -\tfrac{1}{n_2}(1), 0, \ldots, 0]^t, [0, \tfrac{1}{n_2}(1), -\tfrac{1}{n_3}(1), 0, \ldots, 0]^t, \ldots, [0, \ldots, \tfrac{1}{n_{k-1}}(1), -\tfrac{1}{n_k}(1)]^t$$

whose sums of coordinates are each zero. Thus $k - 1$ copies of p_j contribute directly to the join like the non-maximal eigenvalues in (i) and the result follows as stated.

Many well-known examples become part of this general pattern:

1.5 Proposition. The eigenvalues of the complete general n-partite graph $K_{m_1, m_2, \ldots, m_n}$ are $0[M - n]$ (where $M = \sum_{j=1}^{n} m_j$) and the n eigenvalues of the matrix

$$\begin{bmatrix} 0 & m_2 & \cdots & m_n \\ m_1 & \ddots & & \vdots \\ \vdots & & \ddots & m_n \\ m_1 & \cdots & m_{n-1} & 0 \end{bmatrix}$$

(Notation: a number in square brackets following an eigenvalue denotes its multiplicity.) Many calculations are aided by the simple lemma:

1.6 Lemma. Let J_n denote the $n \times n$ matrix whose entries are each 1, and I_n the identity matrix. Then the eigenvalues of the $n \times n$ matrix $sJ_n + tI_n$ are $t[n - 1]$ and $ns + t$.

1.7 Proposition. The complete regular n-partite graph $K_{\underbrace{k, \ldots, k}_{n\text{-fold}}}$ has eigenvalues $(n - 1)k$, $0[(k - 1)n]$, $-k[n - 1]$.

Proof. The matrix in 1.5 becomes in this case $k(J_n - I_n)$ and 1.6 applies to give the result.

Another well-known graph (see [1]) is expressible as a join:

1.8 Proposition. The n-dimensional octahedron is the n-fold join $K_{2, \ldots, 2}$ of the discrete graph S^0 consisting of two vertices.

2. Unary operations on graphs

The binary case of 1.4 gives the largest and smallest eigenvalues of $G_1 * G_2$ as the roots of $x^2 - (p_1 + p_2)x + p_1p_2 - m_1m_2$.

Fixing G_2 in various ways gives unary operations which coordinate and generalise some well-known examples:

We define the cone operation $C : G \mapsto C(G)$ to form the join of G to one point.

2.1 Proposition. If G is row-regular with n vertices of valency p and eigenvalues $\{p, \lambda_2, \ldots, \lambda_n\}$, then the eigenvalues of $C(G)$ are $\lambda_2, \ldots, \lambda_n$ and the roots of the equation $x^2 - px - n = 0$.

The example in 1.2 is such a cone. So are the following:

(i) $K_n = C(K_{n-1})$ for any $n > 1$.

(ii) The star $K_{1,n}$ is the cone on a discrete graph. The latter has eigenvalues $0[n]$, so 2.1 yields the eigenvalues $0[n-1]$ and $\pm\sqrt{n}$.

(iii) The wheel W_{n+1} (see [2]) is the cone on the n-cycle C_n. The eigenvalues of C_n are $2\cos\frac{2k\pi}{n}$, $k = 0, \ldots, n-1$, so W_{n+1} has eigenvalues $2\cos\frac{2k\pi}{n}$, $k = 1, \ldots, n-1$, and the roots of $x^2 - 2x - n$, which are $1 \pm \sqrt{(1+n)}$.

2.2 Proposition. The operation $kL : G \mapsto kL(G)$ of adjoining k loops at each vertex of G has the effect of increasing each eigenvalue of k.

The operation $c : G \mapsto G_c$ of complementation has the following effect on eigenvalues:

2.3 Proposition. If G is regular (not necessarily connected) with n vertices of valency p and eigenvalues $\{p, \lambda_2, \ldots, \lambda_n\}$ then G_c is regular with n vertices of valency $n - p - 1$ and eigenvalues $\{n - p - 1, -1 - \lambda_n, \ldots, -1 - \lambda_2\}$.

3. Row-regularisation of a graph

Row-regular graphs share many of the advantages which regular graphs have over irregular graphs in general. We conclude with a unary operation $\rho : G \mapsto G^\rho$ which 'makes irregular graphs regular', by adjoining sufficient loops at the n vertices of G to make the row-sums in the adjacency matrix all equal to $n - 1$. This row-regularisation operation ρ is well defined, and eigenvalues of graphs of the form G^ρ

are well behaved and give much information about G (see [3]). For example the formula for enumeration of spanning trees in regular graphs given by Wilson [4, 2.11] does not work for irregular graphs, but it does work for row-regular graphs. The operation ρ does not affect spanning trees, and we obtain

3.1 Theorem. If G is an arbitrary graph with n vertices, then the number of spanning trees in G is equal to $\frac{1}{n}\prod_{j=2}^{n}(n-1-\lambda_j)$, where $\{n-1, \lambda_2, \ldots, \lambda_n\}$ are the eigenvalues of G^{ρ}.

References

1. L. Collatz and U. Sinogowitz. Spektren endlicher Grafen. Abh. Math. Sem. Univ. Hamburg, 21 (1957), 63-77.
2. F. Harary, P. O'Neil, R. D. Read and A. J. Schwenk. The number of trees in a wheel. In Combinatorics (D. J. A. Welsh and D. R. Woodall, editors) I. M. A. (1973), 155-63.
3. D. A. Waller. Regular eigenvalues of graphs and the enumeration of spanning trees. Proceedings of Colloquio Internazionale sulle Teorie Combinatorie, Rome (1973).
4. R. J. Wilson. On the adjacency matrix of a graph. In Combinatorics (D. J. A. Welsh and D. R. Woodall, editors) I. M. A. (1973), 295-321.

University College,
Swansea, Wales

GRAPH THEORY AND ALGEBRAIC TOPOLOGY

ROBIN J. WILSON

From a topologist's point of view, the subject of graph theory has little to offer, being essentially a study of one-dimensional objects. To many graph theorists (including myself, until fairly recently), the subject of algebraic topology has seemed somewhat abstruse, and of little interest in graph-theoretical problems. Recently, however, a new topology book has appeared which is considerably more geometrical in content than many of its predecessors, and which uses graphs at various stages to exemplify various theorems and constructions in the text.

The book concerned is C. T. C. Wall, A geometric introduction to topology (Addison-Wesley, 1972).

Our primary aim here is to summarize - from a graph theorist's point of view - some of the aspects of graph theory which are mentioned in this excellent book. In particular, we shall define the cohomology groups $H^0(G)$ and $H^1(G)$ of a graph G, and also the homology groups $H_0(G)$ and $\tilde{H}_0(G)$. We shall use properties of the Mayer-Vietoris sequence to deduce results about $H^1(G)$, and will show how Euler's theorem on plane graphs is a consequence of the Alexander duality theorem.

However, we have a secondary aim, which is an educational one. We believe that pedagogically there is a strong case to be made for using graphs and their properties as motivation for abstract topological ideas, and not simply as examples; for instance, a student who has seen H^1 defined for graphs and calculated using exact sequences is likely to have a much clearer understanding of the topological ideas involved, than one who is first introduced to the concept of a Mayer-Vietoris sequence in its full generality. We shall return to this point later.

Before proceeding, I should like to thank Meurig Beynon and Raymond Flood for several helpful conversations. Any graph-theoretic terms not defined here may be found in [2].

1. Elementary ideas

A graph G consists of a non-empty finite set V of points in R^3 (called vertices), and a finite set E of subsets of R^3, called edges. Each edge e in E is to be homeomorphic to the closed interval [0, 1], where the points of e corresponding to 0 and 1 are vertices in V, and the points of e corresponding to the open interval (0, 1) are points of $R^3 - V$ (see the diagram). We shall give G the topology induced by the usual topology for R^3.

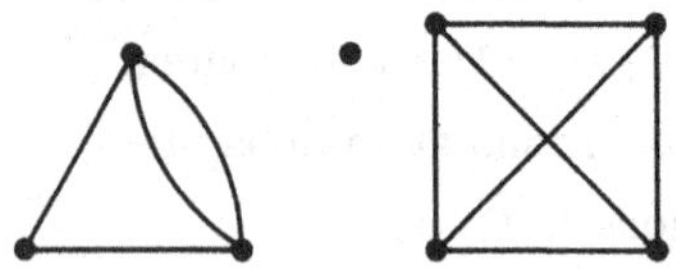

With the topology just described, it is clear that each edge of G is a compact set; since the union of finitely many compact sets is always compact, it follows immediately that every graph G is compact.

We can also discuss the connectedness of G. It is easy to see that a graph is not necessarily a connected space; in general, it splits into a finite number of connected pieces which we call components, and these components are clearly both open and closed in G. It is also a simple matter to see that each component of G is path-connected, since if $P = e_1, e_2, \ldots, e_m$ is a (simple) path (in the sense of graph theory), then we can find a continuous function from [0, 1] onto P, taking the interval $[\frac{i-1}{m}, \frac{i}{m}]$ to the edge e_i, for each i. It follows that G is locally path-connected, and hence - since a topological space which is both connected and locally path-connected must also be path-connected - that the path-components of G must coincide with the connected components.

In a general topological space X, it is usual to define $\pi_0(X)$ to be the set of equivalence classes obtained by taking $x \sim y$ to mean that x is joined by a path to y. It follows from the above discussion that if G is a graph, then $\pi_0(G)$ is simply the set of connected components of G.

2. $H^0(G)$, $H_0(G)$ and $\tilde{H}_0(G)$

If X is any topological space, then the zero-th cohomology group $H^0(X)$ is defined to be the additive abelian group whose underlying set is the set of all continuous functions from X to the group Z of integers endowed with the discrete topology; the addition operation in $H^0(X)$ is performed in the obvious way. The importance of H^0 arises from the fact that if X and Y are topological spaces, then any continuous function f from X to Y induces a group homomorphism $H^0(f)$ from $H^0(Y)$ to $H^0(X)$, under which any continuous function ϕ from Y to Z is mapped to the composite function $\phi \circ f$ from X to Z; in other words, H^0 is a contravariant functor from the category of topological spaces to that of abelian groups.

If G is a graph, then it follows from the discreteness of Z that any continuous function from G to Z must be constant on each of the components of G. It follows that $H^0(G)$ is isomorphic to the additive group of all functions from $\pi_0(G)$ to Z; in fact, there is a corresponding result for any locally-connected space. Note that if G has k components, then $H^0(G)$ is isomorphic to Z^k, the free abelian group on k generators (i. e. the direct product of k copies of Z).

The zero-th homology group $H_0(X)$ of a topological space X is defined to be the free abelian group whose set of generators is the set $\pi_0(X)$. As with cohomology, H_0 behaves functorially, except that this time it is a covariant functor - in other words, if f is any continuous function from X to a topological space Y, then f induces a group homomorphism $H_0(f)$ from $H_0(X)$ to $H_0(Y)$ in the obvious way. It follows immediately from the definition that if G is a graph with k components, then these components are the generators of $H_0(G)$, so that $H_0(G)$ is also isomorphic to Z^k.

We shall also need the reduced homology group $\tilde{H}_0(X)$. This is simply a technical device designed to make the duality theorem go through, and it may be defined as follows: let f be the function from X to a single point c: then, by the above discussion, f induces a function $H_0(f)$ from $H_0(X)$ to $H_0(\{c\})$ (that is, from $H_0(X)$ to Z). We define $\tilde{H}_0(X)$ to be the kernel of this induced function $H_0(f)$. It follows that if

G is a graph with k components, then $\tilde{H}_0(G)$ must be isomorphic to the free abelian group on $k - 1$ generators.

3. The definition of $H^1(G)$

If X is any topological space, we can consider the set of all continuous functions from X to S^1 (the unit circle $|z| = 1$). We shall be concerned with whether or not these functions can be continuously deformed into one another, and with this in mind we define two such functions f and g to be homotopic if there is a continuous function F from $X \times [0, 1]$ to S^1, such that $F(x, 0) = f(x)$ and $F(x, 1) = g(x)$, for all x in X.

It is not difficult to show that the relation 'is homotopic to' is an equivalence relation on the set of all continuous functions from X to S^1, and the equivalence classes are then called homotopy classes. We can define a group structure on the set of homotopy classes by defining the product of the classes containing the functions f and g to be the class containing the (pointwise) product of f and g. This is well-defined, and the resulting abelian group is called the first cohomology group $H^1(X)$.

At first sight, it may seem difficult to see how this relates to graphs. Before looking at the general problem, let us consider two special cases:

Example 1. C_n (the circuit graph on n vertices).

For this graph, we observe that C_n is homeomorphic to S^1, and hence we must investigate the homotopy classes of continuous functions from S^1 to S^1. In fact, as is well known (and can be easily proved), the maps $z \to z^m$ for different integers m lie in different homotopy classes, and all of the homotopy classes can be obtained in this way. It follows that the homotopy classes can be labelled with the integers, and hence that $H^1(C_n)$ is isomorphic to Z.

Example 2. T (a tree).

Since T contains no circuits, there can be no non-trivial continuous functions from T to S^1. It follows that $H^1(T)$ is simply $\{0\}$; we say that T is contractible.

It seems reasonable to guess that if G contains r independent circuits, then each of these circuits can 'wrap itself around S^1' any number of times, so that its first cohomology group ought to be isomorphic to Z^r. This, in fact, turns out to be true, and may be proved in a variety of ways. We shall derive it using properties of the Mayer-Vietoris sequence.

4. The Mayer-Vietoris sequence

A chain of groups G_i and homomorphisms f_i, such as

$$G_0 \xrightarrow{f_1} G_1 \xrightarrow{f_2} G_2 \xrightarrow{f_3} \dots \xrightarrow{f_n} G_n,$$

is called an <u>exact sequence</u> if the image of each homomorphism f_i $(i = 1, 2, \dots, n-1)$ is equal to the kernel of its successor f_{i+1}. If, in addition, G_0 and G_n are trivial, and the remaining groups G_i are all free abelian groups (of rank α_i, say), then it is a simple matter to prove that

$$\alpha_1 - \alpha_2 + \alpha_3 - \dots + (-1)^n \alpha_{n-1} = 0.$$

Of particular relevance for our purposes is an important exact sequence which relates properties of $H^0(X)$ to those of $H^1(X)$. More precisely, if S and T are closed subsets of a topological space X, and $S \cup T = X$, then there is an exact sequence, known as the <u>Mayer-Vietoris sequence</u>, which takes the following form:

$$0 \to H^0(X) \to H^0(S) \oplus H^0(T) \to H^0(S \cap T) \to H^1(X) \to H^1(S) \oplus H^1(T) \to H^1(S \cap T).$$

We shall use the properties of this sequence to derive the following result:

Theorem. <u>If</u> G <u>is a connected graph with</u> n <u>vertices and</u> m <u>edges, then</u> $H^1(G)$ <u>is isomorphic to</u> Z^{m-n+1}.

Proof. If G is a tree, then the result follows immediately from the previous section. If G is not a tree, then G contains an edge $e = \{v, w\}$ which is not an isthmus. It follows, since e is closed in G, that there is an exact sequence of the following form:

$$0 \to H^0(G) \to H^0(G-e) \oplus H^0(e) \to H^0(\{v, w\}) \to H^1(G) \to H^1(G-e) \oplus H^1(e) \to H^1(\{v, w\}),$$

and this sequence reduces to

$$0 \to Z \to Z^2 \to Z^2 \to H^1(G) \to H^1(G - e) \to 0.$$

Let us now assume inductively that $H^1(G - e)$ is isomorphic to $Z^{(m-1)-n+1} (= Z^{m-n})$, and let us also assume (as can be easily proved using induction) that $H^1(G)$ is a free abelian group (whose rank is α, say); it then follows from the above-mentioned property of exact sequences that

$$1 - 2 + 2 - \alpha + (m - n) = 0,$$

giving

$$\alpha = m - n + 1, \text{ as required.}$$

Corollary. <u>If</u> G <u>is a graph with</u> n <u>vertices,</u> m <u>edges and</u> k <u>components, then</u> $H^1(G)$ <u>is isomorphic to</u> Z^{m-n+k}.

Proof. It follows immediately from the definition of $H^1(X)$, that if X can be expressed as the union of two disjoint topological spaces Y_1 and Y_2, then $H^1(X)$ is isomorphic to the direct product of $H^1(Y_1)$ and $H^1(Y_2)$. The result now follows by applying the above theorem to each of the components of G.

(It is worth remarking at this stage that there are various other ways of deriving this result from the Mayer-Vietoris sequence. The reader is referred to Wall's book (Theorem 8. 4 and Exercise 8-25) for further details.)

We can now justify the statement we made at the end of the previous section. If we look at the vector space over the field of two elements, whose elements are subsets of the edge-set of a graph G, and if we add two such sets of edges by taking their symmetric difference, then it is well known (see [1]) that the set of all edge-disjoint unions of circuits of G is a subspace W of this vector space. It is also well known that the fundamental circuits associated with any spanning forest of G form a basis for this subspace W, and it follows immediately from this that the dimension of W (i. e. the maximum size of a set of independent circuits) is equal to m - n + k, as required.

5. Further properties of $H^1(G)$

In this section we shall be discussing the idea of homotopy equivalence, with particular reference to conditions under which two connected graphs are equivalent.

If X and Y are two topological spaces, then we say that X and Y are homotopically equivalent if there are maps $f : X \to Y$ and $g : Y \to X$ with the property that $f \circ g$ and $g \circ f$ are each homotopic to the corresponding identity map. For example, if G is any graph and H is a graph obtained from G by inserting a vertex u of degree two into one of its edges (see the diagram), then G and H are homotopically equivalent, since we may simply map the edge {v, w} into the path {v, u}, {u, w}, and vice versa. As an extension of this, it

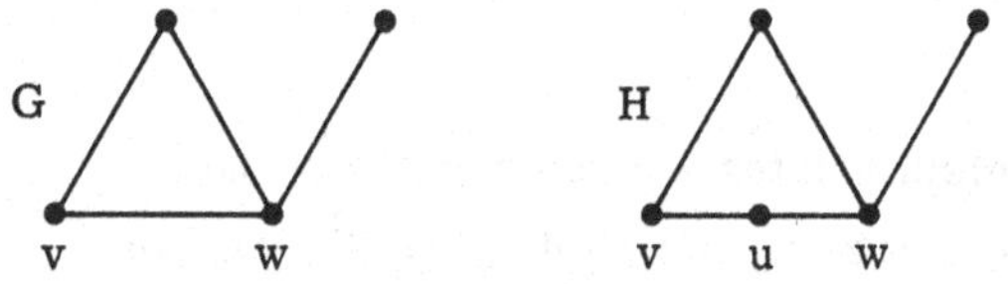

follows that if two graphs are homeomorphic (in the sense of graph theory), then they are homotopically equivalent.

But we can say more than this. Suppose that G is a connected graph, and that H is any graph obtained from G by contracting an edge e (see the diagram); then G and H are also homotopically equivalent,

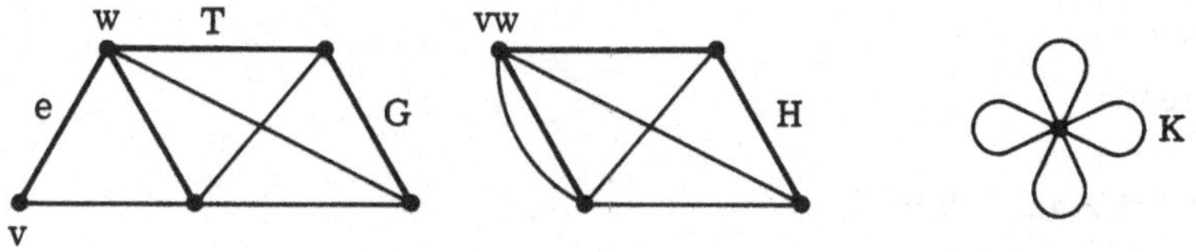

as the reader can easily show. If we now take a spanning tree T of G and contract all of its edges, one at a time, then the resulting graph will be a 'rose' K consisting of m - n + 1 petals. It follows from this that G is homotopically equivalent to K, and we can immediately deduce the following theorem:

Theorem. If G and G' are connected graphs, then G and G' are homotopically equivalent if and only if they have the same first cohomology group.

6. The Alexander duality theorem

If X is a compact topological space which is contained in C, the complex plane, then there is an important relationship between the topological properties of X and those of its complement C - X. This relationship is a special case of a general theorem known as the Alexander duality theorem, and we can state this special case as follows (for a proof, see Wall's book):

Theorem. If X is a compact subset of C, then

$$\tilde{H}_0(C - X) \approx H^1(X).$$

(Note that this isomorphism relates a homology group and a cohomology group. Notice also that if we substitute $X = S^1$, we can easily deduce the Jordan curve theorem.)

We shall now see what happens if we apply this theorem to the case in which X is a graph G. Since X must be a subset of C, we take G to be a plane graph with f faces, n vertices, m edges and k components; in particular, C - G will consist simply of the f faces of G. It now follows from the duality theorem that the rank of $\tilde{H}_0(C - G)$, which is f - 1, must be equal to the rank of $H^1(G)$, which is m - n + k.

Hence

$$n + f = m + (k + 1),$$

which is Euler's theorem for a plane graph with k components.

This may seem a little less surprising if we consider Euler's theorem from an algebraic point of view. If G is a plane graph (which, for convenience, we shall assume to be a polyhedral graph), then any circuit in G may be expressed as the symmetric difference of the boundaries of the faces contained in the interior of that circuit. It follows that the boundaries of the finite faces of G form a basis for the subspace W defined in Section 4, and hence that the number of independent circuits in G (which we know to be $m - n + k$) is equal to $f - 1$, the number of finite faces of G. Euler's theorem then follows as before.

It would be interesting to see whether one could obtain the corresponding result for graphs of genus g ($g \geq 1$) from the Poincaré duality theorem, since the Alexander duality theorem would be inappropriate in this case.

7. Postscript

We should like to conclude by taking up the remarks made at the end of the introduction. If one teaches a course in algebraic topology, it is common to find that students do not react to such theorems as the Alexander duality theorem or to the statement of the Mayer-Vietoris sequence. In the former case, there seems to be an argument for introducing the theory of planar graphs, and the algebraic proof of Euler's theorem in particular, as preparatory motivation for this theorem. In the same way, a discussion of trees, fundamental circuits and rank could well make the statement of the Mayer-Vietoris sequence seem a lot more natural. At any rate, there definitely seems to be a case for introducing a certain amount of graph theory and other combinatorial ideas before plunging in to an algebraic topology course in its full abstraction.

References

1. C. L. Liu. Introduction to combinatorial mathematics. McGraw-Hill (1968).
2. R. J. Wilson. Introduction to graph theory. Oliver and Boyd, Edinburgh (1972).

The Open University,
Milton Keynes, England

APPLICATIONS OF POLYMATROIDS AND LINEAR PROGRAMMING TO TRANSVERSALS AND GRAPHS

D. R. WOODALL

1. Definitions

If $S = \{e_1, \ldots, e_N\}$ is a finite set, $\mathbf{V}(S)$ denotes the set of non-negative vectors on S, i.e., vectors $\mathbf{v} = (v(e_1), \ldots, v(e_N))$ whose coordinates $v(e_1), \ldots, v(e_N)$ are non-negative real numbers. If $\mathbf{v}$ and $\mathbf{w} \in \mathbf{V}(S)$, $\mathbf{v}$ is a subvector of $\mathbf{w}$, $\mathbf{v} \leq \mathbf{w}$, if $v(e) \leq w(e)$ for each $e \in S$. Clearly $(\mathbf{V}(S), \leq)$ is a partially ordered set. If $X \subseteq S$, we usually identify the set X with its characteristic vector $\mathbf{X}$, and write X indifferently in either sense. Thus $\mathbf{P}(S)$, the power set of S, becomes a subset of $\mathbf{V}(S)$, and, if $\phi \leq \mathbf{v} \leq S$, we call $\mathbf{v}$ a subvector of S, writing simply $\mathbf{v} \leq S$. If $\mathbf{v} \in \mathbf{V}(S)$, the cardinality of $\mathbf{v}$ is $|\mathbf{v}| := \sum_{e \in S} v(e)$. The dot product of $\mathbf{v}$ and $\mathbf{w}$ is $\mathbf{v}.\mathbf{w} := \sum_{e \in S} v(e)w(e)$. Thus $|\mathbf{v}| = \mathbf{v}.S$, and if $X \subseteq S$ then $\mathbf{v}.X = \sum_{e \in X} v(e)$. The sum $\mathbf{v} + \mathbf{w}$, intersection $\mathbf{v} \cap \mathbf{w}$ and union $\mathbf{v} \cup \mathbf{w}$ of $\mathbf{v}$ and $\mathbf{w}$ have coordinates $v(e) + w(e)$, $\min(v(e), w(e))$ and $\max(v(e), w(e))$, for each $e \in S$, respectively.

If $(T, \leq)$ is a partially ordered set, I call a subset Q of T subclusive (literally 'closed (or closing) below') if it satisfies the implication

$$x \leq y \text{ and } y \in Q \Rightarrow x \in Q.$$

A subclusive collection of sets is a collection that is closed under the operation of taking subsets (i.e., one that is subclusive in the Boolean lattice). A subclusive collection of non-negative vectors on a set S is a collection that is closed under the operation of taking subvectors.

A matroid $\mathbf{M} = (S, \mathbf{I})$ is a finite set S together with a non-empty subclusive collection $\mathbf{I}$ of subsets of S, called independent sets, such that, for each $X \subseteq S$, every maximal independent subset of X has the same cardinality $r(X)$, called the rank of X.

A polymatroid $\mathbf{P} = (S, P)$ is a finite set S together with a bounded non-empty subclusive collection P of non-negative vectors on S, called independent vectors, such that, for each $\mathbf{v} \in \mathbf{V}(S)$, every maximal independent subvector of $\mathbf{v}$ has the same cardinality $cr(\mathbf{v})$. So

$$cr(\mathbf{v}) = \max\{\,|\mathbf{w}| : \mathbf{w} \in P \text{ and } \mathbf{w} \leq \mathbf{v}\}.$$

Let

$$fr(\mathbf{v}) := \max\{\mathbf{v}.\mathbf{w} : \mathbf{w} \in P\}.$$

Then cr and fr are both non-negative, non-decreasing subadditive submodular functions, but cr is bounded and fr is not. The analogous functions for matroids are both equal to the rank function; so I call cr and fr the confined and free rank functions, respectively. It can be shown that

$$P = \{\mathbf{v} \in \mathbf{V}(S) : cr(\mathbf{v}) = |\mathbf{v}|\},$$

and

$$P = \{\mathbf{v} \in \mathbf{V}(S) : \mathbf{v}.X \leq fr(X) \text{ for each } X \subseteq S\}.$$

(P is a convex polytope: hence the possibility of using linear programming.)

2. The common-base problem

One of the major unsolved problems on matroids is the common-base problem: to find, for each $l \geq 2$ and $r \geq 0$, a necessary and sufficient condition for l matroids defined on the same set to have a common independent set of rank r (or, equivalently, for these same l matroids truncated at rank r to have a common base). Two of the major unsolved problems of combinatorics are of this form, and solutions of them would immediately follow from a solution of the common-base problem. The first is the three-family transversal problem: to find a necessary and sufficient condition for three families of sets to have a common transversal. The second is the Hamiltonian circuit problem: to find a necessary and sufficient condition for a graph (directed or undirected) to have a Hamiltonian circuit. Edmonds [3] (1970) solved the common-base problem for two matroids, yielding a proof of Ford and Fulkerson's two-family transversal theorem [4] (1962).

Theorem 1. Common-base theorem for two matroids. (Edmonds.) Two matroids (S, r_1) and (S, r_2) have a common independent set of cardinality r if and only if, for every partition $S = S_1 \cup S_2$, we have

$$r_1(S_1) + r_2(S_2) \geq r.$$

If $I := \{1, \ldots, n\}$, and $\mathcal{A}(I) = (A_1, \ldots, A_n)$ is a family of sets, let $A(J) := \bigcup_{i \in J} A_i$ for each $J \subseteq I$.

Theorem 2. Two-family transversal theorem. (Ford and Fulkerson.) Two families $\mathcal{A}_1(I)$ and $\mathcal{A}_2(I)$ of sets have a common transversal if and only if, for each pair of subsets J_1 and J_2 of I, we have

$$|A_1(J_1) \cap A_2(J_2)| \geq |J_1| + |J_2| - n.$$

The analogues of these results for three matroids and three families of sets are false. But, using the duality theorem of linear programming, I can prove the following vector results.

Theorem 3. Common-base theorem for polymatroids. Let $(S, fr_1), \ldots, (S, fr_l)$ be l polymatroids, and r a non-negative real number. Consider the following four statements (deleting statement (i) whenever it is not meaningful).

(i) The matroids $(S, fr_1), \ldots, (S, fr_l)$ have a common independent set of cardinality r.

(ii) The polymatroids $(S, fr_1), \ldots, (S, fr_l)$ have a common independent vector of cardinality r.

(iii) For each set of l vectors $\mathbf{v}_1, \ldots, \mathbf{v}_l \in \mathbf{V}(S)$ such that $\sum_{k=1}^{l} \mathbf{v}_k = S$, we have $\sum_{k=1}^{l} fr_k(\mathbf{v}_k) \geq r$.

(iv) For each partition of S into l disjoint subsets $S_1, \ldots, S_l$, we have $\sum_{k=1}^{l} fr_k(S_k) \geq r$.

Then

$$\left.\begin{array}{ll} & (i) \Longleftrightarrow (ii) \Longleftrightarrow (iii) \Longleftrightarrow (iv) \text{ if } l = 2 \\ \text{and} & \\ & (i) \underset{\not\Leftarrow}{\Rightarrow} (ii) \Longleftrightarrow (iii) \underset{\not\Leftarrow}{\Rightarrow} (iv) \text{ if } l \geq 3. \end{array}\right\} \qquad (1)$$

If $\mathbf{v} \leq S$, $\mathbf{v}$ is a vector transversal of $\mathcal{A}(I)$ if $\mathbf{v} = \mathbf{v}_1 + \ldots + \mathbf{v}_n$ where, for each i, $\mathbf{v}_i \leq A_i$ and $\mathbf{v}_i . A_i = \mathbf{v}_i . S = 1$. If, instead, $\mathbf{v}_i . A_i = \mathbf{v}_i . S \leq 1$ for each i, then $\mathbf{v}$ is a partial vector transversal. If $\mathbf{w} \leq I$, let $A(\mathbf{w})$ be the subvector of S whose e-coordinate is equal to $\max\{w(i) : e \in A_i\}$. Using the fact that the partial vector transversals of $\mathcal{A}(I)$ form the independent vectors of a polymatroid (whose free rank function agrees on sets with the rank function of the ordinary transversal matroid), I can derive the following theorem from Theorem 3.

Theorem 4. l-family vector transversal theorem. The following four statements are related by the implications in (1).

(i) The l families $\mathcal{A}_1(I), \ldots, \mathcal{A}_l(I)$ have a common transversal.

(ii) The l families $\mathcal{A}_1(I), \ldots, \mathcal{A}_l(I)$ have a common vector transversal.

(iii) For each set of l subvectors $\mathbf{w}_1, \ldots, \mathbf{w}_l$ of I, we have

$$|S| - \left|S \cap \sum_{k=1}^{l} (S - A_k(\mathbf{w}_k))\right| \geq \sum_{k=1}^{l} |\mathbf{w}_k| - (l - 1)n . \qquad (2)$$

(iv) For each set of l subsets $J_1, \ldots, J_l$ of I, we have

$$\left| \bigcap_{k=1}^{l} A_k(J_k) \right| \geq \sum_{k=1}^{l} |J_k| - (l - 1)n. \qquad (3)$$

(Note that (2) = (3) if $\mathbf{w}_k = J_k$ for each k.)

3. Extensions of Brualdi's theorems

Let G be a finite directed graph, with vertex set V(G). Brualdi proved the following two theorems in 1971, in [1] and [2] respectively.

Theorem 5. Matroid induction theorem. Let $\mathbf{M} = (V(G), \mathbf{I})$ be a matroid. Let

$$\mathbf{I}' := \{S \subseteq V(G) : S \text{ is linked, by some set of } |S| \text{ disjoint paths in } G, \text{ onto a set } T \in \mathbf{I} \text{ with } |S| = |T|\}.$$

Then $(V(G), \mathbf{I}')$ is a matroid.

Theorem 6. Matroid linking theorem. Let $\mathbf{M}_1 = (V(G), \mathbf{I}_1, r_1)$ and $\mathbf{M}_2 = (V(G), \mathbf{I}_2, r_2)$ be matroids, and r a non-negative integer. Then there exist sets $S \in \mathbf{I}_1$ and $T \in \mathbf{I}_2$, with $|S| = |T| = r$, such that S is linked onto T by r disjoint paths in G, if and only if

$$r_2(X) + |Z| + r_1(Y) \geq r$$

for every partition $V(G) = X \cup Z \cup Y$ such that Z separates Y from X in G.

What is remarkable about these theorems is the large number of other theorems that follow from them. For example, the common-base theorem for two matroids follows from Theorem 6 if we remove all the edges of the graph. Menger's theorem also follows from Theorem 6 by a suitable choice of $\mathbf{M}_1$ and $\mathbf{M}_2$. Pym's linkage theorem follows from Theorem 5 and Menger's theorem. Theorem 6 yields practically all results in finite transversal theory, often in several different ways. Theorem 5 yields the result that the partial transversals of a family of sets form an independence structure; Theorem 6 yields its rank function. Theorem 5 yields the fact that the sum of several matroids is a matroid; Theorem 6 yields its rank function. And so on.

It is particularly illuminating to consider the different ways in which (for example) the two-family transversal theorem (Theorem 2 above) can be derived from Theorem 6. Let the two families be $\mathfrak{A}(I)$ and $\mathfrak{B}(J)$, where $|I| = |J| = n$, and let $A(I) \cup B(J) =: S$. Without loss of generality we can suppose that I, J and S are pairwise disjoint. Let G be the graph with $V(G) := I \cup S \cup J$ in which $i \in I$ is joined to $s \in S$ if $s \in A_i$, and $s \in S$ is joined to $j \in J$ if $s \in B_j$. If (to prove Theorem 2) we apply Theorem 6 to G with $\mathbf{M}_1$ the free matroid on I (i.e., the matroid in which all subsets of I are independent) and M_2 the free matroid on J, we are effectively using Menger's theorem. If we take M_1 to be the free matroid on I and M_2 the transversal matroid of $\mathcal{B}(J)$ on S, we are effectively using Rado's theorem on independent transversals. And if we take M_1 to be the transversal matroid of $\mathcal{A}(I)$ on S and M_2 the transversal matroid of $\mathcal{B}(J)$ on S, then the edges of the graph are redundant and we are effectively using the common-base theorem. Thus the matroid linking theorem, applied to the same graph in three different ways, mimics three different proofs of the two-family transversal theorem. It thus goes a long way towards explaining why these three different proofs all work.

Now, the other main proof of Theorem 2 (apart from apparently ad hoc set-theoretic proofs) uses the duality theorem of linear programming. It is thus natural to explore the relationship between the duality theorem and the matroid linking theorem. Is there a single theorem, containing both of these results, that can be applied in different ways so as to mimic even more proofs of Theorem 2? I suggest that, on the contrary, the duality theorem of linear programming is already stronger than the matroid linking theorem. In fact, I can derive both of Brualdi's theorems (via their polymatroid analogues) from the duality theorem.

References

1. R. A. Brualdi. Induced matroids. Proc. Amer. Math. Soc., 29 (1971), 213-221.
2. R. A. Brualdi. Menger's theorem and matroids. J. London Math. Soc., (2) 4 (1971), 46-50.
3. J. Edmonds. Submodular functions, matroids and certain polyhedra. In Combinatorial structures and their applications (Proc. Calgary International Conference, 1969; Gordon and Breach, New York, 1970), 69-87.
4. L. R. Ford and D. R. Fulkerson. Flows in networks. Princeton University Press (1962).

University of Nottingham,
Nottingham, England

PROBLEM SECTION

1. A theorem of Gleason states that a finite projective plane with the property that the diagonal points of any complete quadrangle are collinear, is a Desarguesian plane over a field of characteristic 2. I call a biplane an incidence structure in which any two points are incident with just two blocks, and any two blocks with just two points. A question similar to Gleason's: which biplanes have the property that, given three points in a block, the three blocks incident with two of the given points have a common point? (Fig. 1)

Figure 1

Figure 1 is a picture of one such biplane, with 4 points. There is just one more example known; it has 16 points.

Peter J. Cameron (Oxford)

2. Let E be the real Euclidean plane with the usual distance function d. Prove that, given $\varepsilon > 0$ and $n \geq 6$ points $P_1, P_2, \ldots, P_n$ of E, there exist n points $Q_1, Q_2, \ldots, Q_n$ of E satisfying

(a) $d(P_i, Q_i) < \varepsilon$ for $1 \leq i \leq n$, and

(b) all the distances $d(Q_i, Q_{i+1})$ and $d(Q_i, Q_{i+2})$ are rational.

(Here $1 \leq i \leq n$ and $Q_{n+1} = Q_1$, $Q_{n+2} = Q_2$.)

D. E. Daykin (Reading)

3. Let C_n^l be the graph with vertices $\{1, 2, \ldots, n\}$, with vertex i adjacent to vertices $i-l, \ldots, i-1, i+1, \ldots, i+l$ (mod n) for each

i=1, ..., n. Prove that if G is a graph on n vertices with minimum valency d, and $d \geq n(1 - \frac{1}{l+1})$ for some positive integer l, then G contains C_n^l as a subgraph. The case $l = 1$ is just Dirac's Theorem. It is easy to prove for $l \geq \frac{1}{4}n$, but the general case must be expected to be difficult since it would have, as a corollary, the remarkably difficult Hajnal-Szemerédi Theorem [Proof of a conjecture of P. Erdős, Combinatorial Theory and its Applications, pp. 601-23, North Holland, 1970]. This theorem states that if a graph G on n vertices has maximum valency less than r, the G is r-colourable such that the sizes of the colour classes are all $[n/r]$ or $\{n/r\}$.

P. D. Seymour (Oxford)

4. Let $c > 0$. Suppose we are given a sequence $S_1, \ldots, S_n$ of n non-empty sets of disjoint intervals in the real line, such that each interval in each set overlaps at least one interval in each preceding set. Then does there necessarily exist a point in at least cn of the intervals? I conjecture that this is so when $c = \frac{1}{4}$. (I know that $c = \frac{1}{3}$ is too big.) But at the moment I cannot prove it for any positive value of c. (I have an algorithm for a certain linear storage problem, and I need to be able to prove this (for some $c > 0$) in order to prove that my algorithm works.)

D. R. Woodall (Nottingham)

Notes on problems raised at last year's conference*

1. Last year I raised the question of a possible connection between the 'perfect matching' and the 'tree' problems on the plane square lattice. For a large lattice the number of trees on a lattice of N points is the same as the number of perfect matchings on a lattice of 4N points, that is on a lattice of half the spacing. Is there a (1 - 1) correspondence between the trees and the matchings? You can't draw such a conclusion from the fact that the numbers are asymptotically equal. I have, however, found quite a simple construction that proves the existence of such a correspondence for finite lattices.

* See Combinatorics (Proc. 1972 Oxford Combinatorial Conf.; ed. D. J. A. Welsh and D. R. Woodall; Institute of Mathematics and its Applications, 1972), 359-63.

Consider the graph 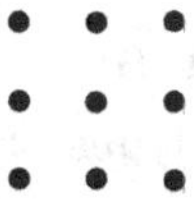 of nine points. We can build up a portion of a plane square lattice by 'butting' these graphs together, deleting rows of three points, e. g. from four such graphs we make the graph:

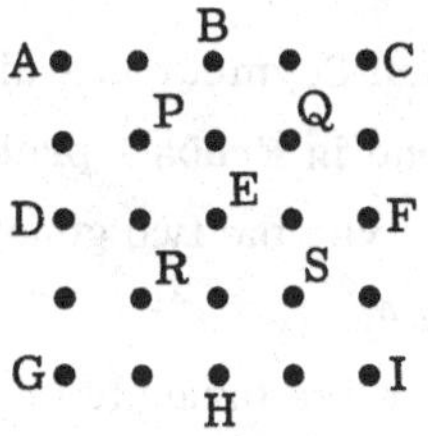

Such a graph always contains an odd number of points. Delete the top right-hand point and the two lines incident to it. I claim that there is a (1 - 1) correspondence between perfect matchings on this graph and trees on the points A B C D E F G H I.

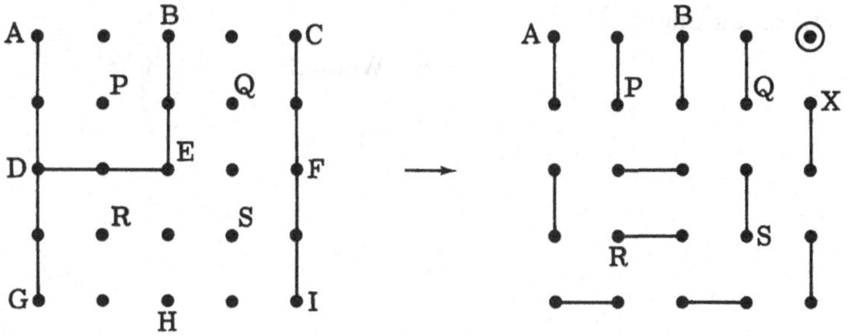

Given the tree, this construction is in two steps.

(1) Match the centres of the squares P, Q, R, S with neighhouring points <u>not</u> on the tree, beginning with 'blind alleys' surrounded by three lines such as P and R. An inductive argument shows that this can be done in just one way.

(2) Starting with the terminal vertices of the tree work inwards along each branch of the tree in turn matching alternate pairs of neighbouring points. We begin with the vertice of type X next to the deleted point C. (There may be one or two such vertices.) We then go on to

the other terminal vertices like A and B.

There are a number of possible types of case but I am fairly confident that I have considered them all, so that the correspondence is indeed (1 - 1). Similar results must exist for the triangular and hexagonal lattices, but I have not looked at them in detail.

H. N. V. Temperley (Swansea)

2. My problems 2, 3 and 5 remain open. Paul Seymour has disproved the conjecture in my problem 1 (and the one in Erdős's problem 2). Following an idea of Seymour's, I have disproved the two generalizations on the thrackle conjecture in my problem 4.

[The present most general possible form of the thrackle conjecture seems to be the following: Let G be a drawing in the plane of a finite graph without loops, multiple edges, self-intersecting edges, or intersecting adjacent edges. Suppose that, if we direct each edge of G arbitrarily, the number of times each edge e_i crosses each edge e_j (not adjacent to e_i) from right to left is one more or less than the number of times it crosses from left to right. Then G has at least as many vertices as edges.]

D. R. Woodall (Nottingham)